# Birds of Bali

Once I was part of the music I heard
   On the boughs or sweet between earth and sky,
   For joy of the beating of wings on high
My heart shot into the breast of a bird.

I hear it now and I see it fly,
   And a life in wrinkles again is stirred,
   My heart shoots into the breast of a bird,
As it will for sheer love till the last long sigh.

GEORGE MEREDITH


Rise up, my love, my fair one, and come away.
For, lo, the winter is past, the rain is over and gone;
The flowers appear on the earth; the time of the
   singing of birds is come, and the voice of the
   turtle is heard in our land.

THE SONG OF SOLOMON

*BALI STARLING*

# BIRDS
## of
# BALI

## Victor Mason

*Illustrations by*

## Frank Jarvis

*With photographs by Morten Strange*
*and endpapers by I Made Budi and his son I Made Moja of Batuan*

PERIPLUS
EDITIONS

**Published by Periplus Editions (HK) Limited**

### DISTRIBUTORS

**Australia:**

*New South Wales:*    R & A Book Agency,
Unit 1, 56-72 John Street Leichhardt 2040

*Northern Territory:*    Channon Enterprises, 8 Davies Street, Jingili NT 0810

*Victoria:*    Ken Pryse & Associates,
156 Collins Street, Melbourne 3000

*South Australia:*    Oriental Publications, 65A South Terrace, Adelaide 5000

*Northern Queensland:*    Queensland Book & Maps,
First Floor, 37 Tully Street, South Townsville 4810

*Southern Queensland:*    Robert Brown & Associates,
67 Holdsworth Street, Coorparoo 4151

*Western Australia:*    Edwards Book Agencies,
Unit 4, 48 May Street, Bayswater 6053

**Benelux:**    Nilsson & Lamm B.V.,
Pampuslaan 212-214, 1382 JS Weesp, The Netherlands

**Hong Kong:**    Asia Publishers Services Ltd., 16/F,
Wing Fat Commercial Building, 218 Aberdeen Main Road

**Indonesia:**    C.V. Java Books,
Cempaka Putih Permai, Blok C-26, Jakarta Pusat 10510

**Singapore/Malaysia:**    Berkeley Books Pte. Ltd.,
2A Paterson Hill, Singapore 0923

**Thailand:**    Asia Books Co. Ltd., 5 Sukhumvit Soi 61,
Bangkok 10110.

For information on distribution in all other areas please contact
Periplus (Singapore) Pte Ltd.
2A Paterson Hill
Singapore 0923

Publisher : Eric M. Oey
Editor : David Pickell
Inside design : David Pickell
Cover design : Peter Ivey

Library of Congress Catalog Card Number: 89-61113
ISBN: 0-945971-04-4

First printing 1989
Second printing 1993
Third printing 1994
Printed in the Republic of Singapore

Cover Illustration by Frank Jarvis. *A Javan Kingfisher in Bali.*
Front endpaper: *Birdwatching in the Monkey Forest*, by I Made Moja, Batuan
Back endpaper: *Birding Map of Bali*, by I Made Budi, Batuan

# Preface

To John and Jonquil,
My companions in the field,
And to Jassy,
My companion at home.

Whoever heard of a preface being composed before the main body of a book? Yet that is precisely what I am doing, because in spite of all appearances, there is a predetermined plan and format, from which I may not deviate one inch. Knowing the way it will turn out, I use this simple and expedient ploy to broach a subject which is ordinarily presented in field guide form, tabular and graphic, and designed for your dedicated bird-watcher; but which is here accorded anything but a conventional or systematic treatment, and is intended for more general consumption by anyone professing the slightest interest in birds, or indeed in anything at all aside from matters mundane.

Of the 300 or so species of birds reported to visit or live in Bali, probably rather less than 100 are likely to be encountered by the casual observer, which is roughly the number described and illustrated in this little book. Many of them may be seen without the need to hire a sailing-boat or motor-car, even to vacate comfy chair and exit through the garden gate.

No attempt is being made to follow the order of species normally adopted by other authors, i.e. starting with sea-birds and ending with finches or crows. If there is to be a theme, it is one of progression, and I think it is logical to introduce the birds in more or less the order in which one is likely to come across them, beginning with birds common or garden and working our way through to those found only in the wilderness. But there are bound to be some unexpected twists and turns in the trail, and the odd challenge may be thrown in for good mea-sure en route. For instance, two little king-fishers are mentioned which must be very rare here, if not extinct. Though both can be seen in Java and elsewhere, neither one to my knowledge has been observed in Bali for a good many years, or if it has, there is no record of it. How wonderful it would be to be reassured that they are *in esse* here, alive and well, and then to declaim it loudly from the rooftops so that all who would hear might know and rejoice.

The selection of species for discussion is somewhat arbitrary. How could it be other-wise? Most of the commoner residents are being included as well as some which are notable for their beauty or rarity. Sea and shore-birds are given short shrift, since nearly all are migratory and can be seen elsewhere, it may be to better advantage; besides which they are well covered in any number of books. Care is being taken to include a fair number of native species (at least 24 plus 7 distinct races) which are not illustrated in Ben King's *Field Guide to the Birds of South-East Asia,* which is nonethe-less indispensable for anyone with an inter-est in the birds of this region.

As a bonus to serious observers a com-prehensive and systematic check-list of all birds recorded as having occurred in Bali is added as an appendix; and it is recommended that copies be made for use in the field.

I shun the use of technicalities, albeit that I am bound on occasion to have recourse to ornithological terms when dis-cussing some finer point of identification; and you will find no glossary in here. Above all, the aim throughout will be to bring the birds to life, for of all God's creatures they are undoubtedly the liveliest. They do not live in reference books or museum collections, and they can barely be said to live much of a life when confined to a cage or aviary, no matter how large or well built.

Perhaps this is not the place to put in a special plea for restraint in the practice of keeping caged birds as pets, but I will do it anyway. Some species are protected by law and may not be trapped and held captive. And they most certainly may not be shot. There is far too much wanton and deliberate destruction of our bird-life. So many vis-

*Photo by Morten Strange*

*Bali Barat National Park (Dry Season)*

itors to Bali have remarked to me on the paucity of birds in general, and whilst this does not hold true of every area, the contention is broadly correct. When I first came to Ubud in 1974, Ashy Drongos and Black-naped Orioles were always present in my neighborhood, and the Serpent Eagle was a transient visitant near daily. The drongos have gone; the others all but vanished. I seldom see a hawk of any kind nowadays — anywhere. It will require courageous justice to impose a ban on the

use of air guns throughout the land.

Now, there comes a time, during even the briefest and hardest earned holiday, when the urge arises to quit the comfort of air-conditioned room, poolside bar, or shopping centre, and head off into the hinterland. For those who inhabit this place, it has always been the custom to turn inward the head and lift up the eyes to the hills, to that seat where gods indeed do dwell. And who has not been possessed by a sudden urge to tread the untrodden track, to scale some dimly beheld forest fastness and venture into the *altiplano inexplorado* in search of the lost city — the unknown?

We are lucky in Bali. There is still a wilderness well removed from man and his agriculture, his industry and all his noisome productions. It is not difficult to attain.

The mountain spine that extends the entire length of the island is clad in dense forest, while the westernmost part harbours a large area of jungle and scrub that reaches to the coast. This latter region, now designated a National Park and nature reserve by the Indonesian Government, is the richest in wild-life, providing a suitable habitat for a tremendous number and variety of birds, including Bali's one true endemic species, variously known as Rothschild's Myna or Grackle, or simply as the Bali Starling, *Leucopsar rothschildi*.

Elsewhere, the only open tracts of land that remain to some extent uncultivated may be found in the east, and on the southernmost tip of the island. The former, a narrow strip running between the mountains and the sea, is an arid zone of volcanic debris supporting little in the way of life, following the spectacular eruption of Gunung Agung in 1963; while the latter comprises a limestone bluff or headland, which must have been thrust up from the ocean bed in comparatively recent times, and which contains a goodly population of birds, notably those favouring dry open country, such as drongos, starlings and shrikes.

For visitors interested to see forest birds but without much time to spare, probably the best bet is to head for the temple, Pura Luhur, situated on the southern slopes of Batukau. Less than one hour's drive from Sanur, going north-west via Tabanan to Wongayagedé, will bring you there. Now the tarmac road leads directly to the temple precincts. Inspect the vast wayside *ficus* trees for marvellous views of less common birds like barbets, minivets, and fruit-doves; then wander through a maze of jungle paths, bordered by hibiscus and ixora, past sacred moss-grown shrines and watering places, where all the flowerpeckers and flycatchers and a host of other birds will appear magically *en masse*, so that all you can do is lower your binoculars and gape in wonderment.

Another good spot for watching jungle birds is an area in the central highlands, where the three lakes — Bratan, Buyan, and Tamblingan — placidly repose beneath towering tangled mountain slopes. The journey to Bedugul takes barely above one hour, and there is a fair choice of accommodations in this hill resort. Several trails lead down along the lake to fruitful bird watching in the nearby jungle. Or simply sit in the rest-house gardens, sipping a cool drink, and observe the hordes of small birds invading the flowering shrubs — honeyeaters, flowerpeckers, sunbirds, white-eyes, and so on. Try to be there on a week-day, the earlier the better, since the birds are not partial to the dinning of the motor-boats.

On the other side of the village are the botanical gardens. Take the small road on the left before reaching the market-place; pass groves of citrus and passion-fruit vines, plots of gladioli, dahlias, and African marigolds, to the entrance gate, where a very modest admission fee will be demanded. On week-days there is seldom anybody there and you will have the entire grounds to yourself. The margin between garden and jungle is often hard to detect; thus splendid opportunities are afforded to view many shy forest birds in the open; for example, ground-thrushes, cuckoo-shrikes, scimitar-babblers, and the glorious Golden Whistler. And if you proceed with the utmost caution, and fate is propitious, you might just get a glimpse of the resplendent Red Junglefowl, *Gallus gallus*. I have heard it there on several occasions — a crow much like that of your barnyard bantam — but never actually seen it. If it is any consolation, there are sometimes one or two birds on display and for sale in the market-place. Like other captive birds there, their condition is disconsolate to say the least.

*Highland forest near Lake Tamblingan*

A couple of miles further on along the main-road is the Bali Handara golf club. The access road and course borders are ideal for bird-watching, as are the stream-bed and pipe-line which may be gained by following the track commencing from the road by the small artificial lake. Here may one have unrivalled views of barbets and ground-thrushes, as well as some tricky warblers: watch out in particular for the Indonesian Bush-Warbler, in damp thickets near the pipe-line.

It is a fact worth pointing out that creatures of the jungle are invariably best observed from roadways and water-ways, and open spaces provided by arboreta and golf-links in a forest setting; whereas interior trails through dense vegetation avail little in the way of visibility, and although one may obtain a brief but rewarding view of some great rarity, on the whole most birds tend to stick to the canopy, where direct sunlight promotes the growth of fruit, nectar and insects on which they feed.

Which having said, there is one expanse of forest in the region which is well worth a visit, and this surrounds the third of the

Photo by Morten Strange

*Benoa Bay mud flats*

three lakes, which is called Tamblingan. Vehicular access is possible by taking a turn off the road to Munduk which overlooks the northern edge of Lake Buyan: but it cannot really be recommended during the rainy season, even with four-wheel drive, while in the dry, a veritable dust-storm is created by everything on wheels, to the acute discomfort of all who happen to pass by. Far better to take the track which skirts the south side of the lake (Buyan), proceeding as far as one can safely go with trans-

port; then walk through the jungle, over the intervening ridge, and down to the shores of Tamblingan. The setting here is magnificent, true primary forest with massive trees and sparse undergrowth, encircling a crescent of cattle pasture and cultivation beside the shimmering lake; two tall temple pagodas survey the scene. Birds are in abundance, and it is one of the richest patches of jungle I have seen. Particularly good are the malkohas, babblers, woodpeckers, and ground-thrushes. Visibility through the lower and middle storeys is perfect.

But undoubtedly the best area is the previously mentioned National Park in West Bali (Bali Barat). To get the most out of the place one should consecrate a few days or more. First, for the sake of good order, an entry permit should be obtained: for this report to the wardens at the P.P.A. (Perlindungan dan Pelestarian Alam) establishment in Cekik, next to the main-road junction one mile before Gilimanuk. Accommodations can be a problem, but there is sometimes a bed available at Labuan Lalang, where the boat puts out to the small island of Menjangan. Highly recommended is the Losmen Margarana, beautifully situated in a coconut plantation near the beach, about half a mile west of the Jayaprana temple at Teluk Terima. And failing that, there are hotels in Gilimanuk, of which the most convenient is Nusantara II, on the cape opposite the ferry wharf.

A great tract of wilderness affords a wide variety of habitat, including jungle, scrub, savannah, seasonal swamp, and mangrove. Here may one see hornbills, pittas, rollers, treeswifts, and other birds not seen elsewhere, not to mention the celebrated Bali Starling. Here also, commencing around mid-October, can one catch the annual migration of the raptors, with really impressive numbers of honey-buzzards and sparrow-hawks, interspersed with a sprinkling of other species, coming over from Java and heading southeast. And should one wish to take a break from the rigours of bird-watching and jungle-bashing, there are some lovely beaches in the vicinity.

Finally, if we have any time left at all, the island of Penida, known also as Bandit Island, beckons on the south-east horizon. Boats may be hired in Sanur, and there is a regular *jukung* ferry service from Kusambe, east of Klungkung. Rather basic lodgings may be had at Sampalan and Toyapakeh; and a complex of hostels patronized by the surfing fraternity has been established on the smaller island of Lembongan. Penida is utterly different from Bali, an arid mass of limestone outcrop, sparsely populated and devoid of surface water, with closer faunal affinities to Lombok and the Wallacean region. An excursion to this other-worldly realm will yield a wealth of ornithological surprizes, including species not found at all on mainland Bali, such as White-faced Heron, Red-chested Flowerpecker, and Black-faced Munia. It is also the best place to see White-tailed Tropicbirds, hawks like Osprey, Peregrine, and White-bellied Sea-Eagle, Imperial Pigeons (mostly shot out now), Common Koels, Mangrove White-eyes, and Hair-crested Drongos.

Having made it clear at the outset that the aim here is to introduce the reader gently to some of our more obvious Bali birds, as well as one or two not so obvious ones, I hope the appetite will be sufficiently whetted for further and more instructive literature on the subject. Ben King's *Field Guide to the Birds of South-East Asia* I have already mentioned, and this is required reading for every birder. *Birds of Java and Bali* by John MacKinnon covers all species, and Collins will shortly publish a *Field Guide to the Birds of Indonesia* (excluding West Irian). There is no better bird book than Bertram Smythies' *Birds of Borneo* (obtainable from The Malayan Nature Society, P.O. Box 750, Kuala Lumpur): nor is there any handier, more concise and comprehensive work than *Birds of Malaysia* by Jean Delacour (now sadly

out of print, but a photo-copy would do).

If I were to thank anyone for stimulating my interest in the birds of Bali, it would be the lively English lady who bounced into my house some years ago and proceeded to bombard me with questions about our local feathered friends, which I was woefully unable to answer. I resolved there and then never to find myself again at such a wretched disadvantage. Above all, I thank John Ash for opening wide my eyes, and for his incredibly generous and good-natured guidance in the field.

And lastly, should this essay be found fit

Photo by Morten Strange

*The coast of Nusa Penida*

not merely to inform but also to entertain, then it will have fulfilled its promise and will not have been in vain. Yet, having regard to the illustrations by Frank Jarvis, I am not at all convinced that one word from me is called for. They are, as you must see, much more than plain illustrations, skilful graphics, or the like. These are portraits, painted on the spot, of living birds in all their glory, set in a timeless sphere. Without them, we should remain, rudderless and disenchanted, in the dark.

# Alphabetical Contents

EMERALD DOVE

PINK-NECKED PIGEON

# Java Kingfisher

*Halcyon cyanoventris*

"What is that incredible blue bird with the huge red beak?" So often is the question posed by visitors to Bali. It is curious that mention of one particular bird should so frequently be made in the general course of conversation. Yet, considering its stunning beauty, it is perhaps not so remarkable that the Java Kingfisher should at once commend itself to the captive eye and make an indelible impression on all who behold it.

If we were to describe the bird accurately in field guide format, we should mention, under the scientific name *Halcyon cyanoventris,* that the length including bill is 11 inches, that the head is dark brown, throat and breast chestnut, belly dark blue, back purple, wings and tail bright blue, bill and feet red. Which is over-simplifying it, but still tedious. And does it really describe the bird? Perhaps we should leave it as the "incredible blue bird with the huge red beak" which sums it up rather well and serves to distinguish it from any other bird one may find in Bali.

However, it certainly calls for closer inspection through a good binocular, in the soft and subtle light of early morning and late afternoon, when all birds are most active and thus more easily observed. Notice how the colours seem to assume a different hue — the belly now a rich royal blue tinged with purple, wash of cobalt on the breast; the mantle now a brilliant Tyrian purple, and wing and tail feathers azure to turquoise in the delicate reflected light. Look again — drat! The bird has flown, in a low straight line across the fields, bright white wing patches starkly revealed in its headlong flight.

Listen to the noise it makes — a shrill, harsh, laughing, staccato call: "*tjie tjie tjie tjie tjie tjie tjie*" — the first note emphatic, the remainder seeming somewhat to diminish in pitch and scale. Also a sort of song, soft and mournfully tremulous, descending the scale: both utterances oft repeated either in flight or at rest.

Small wonder then that the Java Kingfisher impinges so essentially on the senses of even the most casual observer. It is, moreover, endemic in the islands of Java and Bali. Nowhere else may it be found. Here, in Bali, it is by no means uncommon in the lowlands, frequenting the river-beds of verdant ravines as well as the open expanses of *sawah,* where it may often be seen perched atop a slender pole or thatched roof of shrine, sacred to Dewi Sri, goddess of agriculture and fertility. From such vantage the bird swoops down on its unsuspecting prey, consisting in the main of lizards, eels and frogs, and insects of the larger sort. I have one which comes sometimes of a late afternoon to take up station on my verandah rail, pouncing upon the winged male termites as they emerge from under the boards. A valuable service indeed! And as the predator strikes, I am reminded always of Byron's lines on The Destruction of Sennacherib:

> The Assyrian came down like the
>     wolf on the fold,
> And his cohorts were gleaming in
>     purple and gold.

Now we should not be put off by the scientific names which are mostly derived from classic Latin and Greek. First we have the generic or group name which defines the bird's affinity with other members of its group or genus, and then the specific name which serves to identify the bird precisely. In this case *cyanoventris* (from Greek *kuanos,* dark blue, and *venter, ventris,* Latin for belly), and *Halcyon* (from Greek *hals,* the sea, and *kyon,* conceiving) which is an old poetic term for kingfisher. The ancients believed the birds to be a model of constancy and affection, and if the male bird lost its mate, it would choose to remain forever solitary rather than pair again. They also

supposed that the birds conceived and built their nests on the sea which remained calm while the female was brooding and bringing forth her young: hence the expression, halcyon days, which are calm and happy days — kingfisher days.

Being so spectacularly caparisoned, and thus so frequently remarked and talked about; and seeing that it is known only from here and Java, Frank Jarvis and I felt that it was fitting that the Java Kingfisher should be selected to head our list of birds.

# Collared Kingfisher
*Halcyon chloris*
CENTRE

Here we have another halcyon which commonly frequents the coast, and which must sooner or later make its presence both seen and heard by anyone staying at one of the resort hotels. This is the wild, shrieking *chloris*, known also as the White-collared or Mangrove Kingfisher, whose green to peacock-blue upperparts and white underparts render it instantly recognizable. In Bali it seems equally at home in mangrove, garden, rice-field, scrub, or forest, and so we have seen it in every area and at every elevation. Like the shrikes, bee-eaters, and turtle-doves, it may often be observed perched prominently on overhead wires.

# Sacred Kingfisher
*Halcyon sancta*
FLYING

On several occasions have keen bird-watchers mentioned to me that they have seen a Collared Kingfisher which "had something wrong with it" or "definitely didn't look right." What they had in fact seen was this austral winter migrant which is with us from about April to October in fairly large numbers. Closely resembling its near relative in dress though somewhat smaller, the turquoise crown and mantle are much duller, and the white collar and underparts are irregularly marked with buff, as shown on the pictured bird winging its way through the mangrove. In fact it is exactly what one might expect the Collared Kingfisher to look like after having partaken of a mud bath.

Altogether a quieter and more restrained bird than the frantic, raucous *chloris*, which is a further pointer to identification, it may easily outnumber its northern counterpart on the coast, where it favours mangrove and mud in particular, seldom venturing inland.

No fewer than eight species of kingfisher (compared with only one in Europe) have been recorded here, but three of these are extremely rare if not extinct; and they are presented here not only for their striking plumage, but also as a challenge to the adventurous spirit who would seek them out and rejoice in their rediscovery.

# River Kingfisher
*Alcedo atthis*
TOP RIGHT

Strictly classed as a race of the European kingfisher from which it differs not at all, it must be decidedly uncommon here, and might still be found by waterways in open country near the coast. This bird is also called the Common Kingfisher.

# Blue-eared Kingfisher
*Alcedo meninting*
TOP LEFT

Similar to the last species, but slightly smaller, and a much darker, richer blue, it is also much shyer, being found only by streams in jungle and adjacent plantation.

# Stork-billed Kingfisher
*Halcyon capensis*
BOTTOM

Aptly named for its enormous red bill (though I have observed that in immature birds the colour is black), it has been seen but three or four times in recent years in the west of Bali. Its preferred habitat is lightly wooded streams and mangrove, where in spite of its large size and conspicuous plumage, it may well escape detection, for it tends to sit in a shady bower, silent and motionless, for hours and hours on end. Only by its infrequent flurried flights and harsh chuckling calls is the majestic presence finally revealed.

JARVIS

# Small Blue Kingfisher

*Alcedo caerulescens*

Another veritable gem of the animal kingdom — at one instant sapphire, and the next turquoise, depending on the angle of the light — it dazzles the eye of the beholder in Bali and Lombok, where it is locally quite common; and elsewhere it has been recorded only in parts of Java and Sumbawa.

Resembling a miniature version of the well-known screeching *chloris,* this brilliant blue and white bird is to be found in patches along the coast. The best place to see it is perched on the small-fry fishing traps at the southern end of Sanur beach, or streaking like a bullet up and down the network of fishponds between mangrove and mainroad. It is a rewarding experience for anyone prepared to stroll the length of the beach to discover this diminutive treasure, and I heartily recommend it. Few people in the world can claim to have seen the Small Blue Kingfisher in its natural state.

Like other members of the genus, it selects a steep bank in which to excavate a tunnel of two or more feet in length, ending in the nesting chamber where is laid the clutch of four or five eggs, white, perfectly round, and glossy. The only utterance in flight is a high-pitched squeaking peep.

# Rufous-backed Kingfisher

*Ceyx rufidorsus*

Possibly the most beautiful of all its tribe, this living jewel warrants an expedition if it is to be properly appraised. As the Small Blue Kingfisher is the lapis lazuli of the lagoon, so is this tiny bird variously the ruby, jacinth, or amethyst of the forest. It simply depends on the light.

Frank Jarvis likened it to a piece of translucent porcelain through which shines an inner light; and indeed, beheld in the dimness of its customary forest setting, it has a radiance and luminosity which I find hard to describe. Frank's portrait captures it perfectly. And I can think of one other creature in Nature which is also found in Bali, the lovely Malayan Lacewing butterfly, the males of which share the distinction in having the upperparts a resplendent fox-red tinged with lilac. Truly a breathtaking combination of colour.

Exclusively found in forest, a march upstream from the road-bridge at Teluk Terima in the Bali Barat National Park should yield good views of one or more, perched on a bough above a suitable pool, or whizzing through the trees between the reaches, as a rose-red shooting star to illuminate the gloom.

JARVIS

# Yellow-vented Bulbul

*Pycnonotus goiavier*
BOTTOM

I think, without any doubt, the commonest birds in Bali must be these ubiquitous bulbuls of the Yellow-vented variety. You will see them everywhere, usually in smallish parties, in unceasing chorus of chatter and chortles, as they race each other from palm to palm. The Balinese name is *tjerutjuk*, which sums up the noise they make rather well, and their breezy bubbling song is one of the first sounds to greet the early riser. Some writers affect to be irritated by it, finding it tuneless and repetitive. They may not be early risers, or if they are, then perhaps they habitually get out of bed on the wrong side. It is a sound I associate with the gardens of the East, and one that is most typical of my own garden. And how refreshed I always am to hear it on wakening in my own bed, for the first time after journeying to distant lands. How infinitely more refreshing, I reflect, than the babel of human voices or the dinning rattle of the motor-car.

One feature of the Yellow-vented Bulbul which is curiously omitted from most published descriptions is the fairly prominent brown coronal stripe. This is erectile and rather ragged, giving the bird a slightly scruffy, unkempt appearance, as if it could do with a good brush and comb.

# Sooty-headed Bulbul

*Pycnonotus aurigaster*
TOP

Of the six types of bulbul occurring in Bali, one species should not be unfamiliar to residents of the southern coastal resorts, especially Sanur and Nusa Dua. This is the Sooty-headed Bulbul, with blackish and slightly crested head, giving a somewhat domed appearance. The vent, or under tail coverts, evidently varies from red to yellow according to race, but all the birds seen here have vents of a colour I would describe as hot orange. There is a pale patch on the rump which is most distinctive in flight.

The Sooty-headed has a more melodious voice than its Yellow-vented cousin, including an oft-repeated clear whistled triplet which forms a familiar portion of the dawn chorus.

Formerly confined to southern coastal areas, I have noticed a considerable expansion of the range in recent years and have records of breeding in the Ubud area. Since none of the early ornithologists visiting Bali reported the species, we may assume the bird was introduced in very recent times; and since it is a popular caged-bird, the currently expanding population is no doubt descended from a handful of feral refugees which became established in the neighbourhood of Denpasar.

# Orange-spotted Bulbul

*Pycnonotus bimaculatus*
CENTRE, RIGHT

Found only in Sumatra, Java, and Bali, it is a shyer, wilder bird of highland forest, though I have actually encountered it in the cliff-top temple at Ulu Watu. On either side of the forehead are the distinctive orange spots, and the cheeks are a lovely shade of lemon-olive.

The voice, which is most pleasing to the ear, I have described in my notes as a soft and sweet bubbling cadence.

Young bulbuls have the same basic plumage pattern as their parents — brown tinged olive overall, with belly whitish and vent yellow — but without the eponymous spots.

# Bar-winged Prinia
*Prinia familiaris*
CENTRE

It is time to consider some of our smaller garden birds, though what they may lack in size they do amply make up for in their vigour and agility, not to say their clamorous contributions from before sun-up to nightfall. Take for instance those neat little, longish-tailed, olive-grey–brown birds, clambering about in the copper-leaf and croton bushes, every so often emitting a plaintive *twee-wee-wee*, succeeded by utterances of quite explosive force. Notice the long white-tipped tail feathers, white throats and upper breasts, twin white wing bars, fleshy apricot legs, amber eyes, and lemon-yellow bellies. They are the Bar-winged Prinias, or Wren-Warblers, as I prefer to call them, which are equally at home in ornamental garden, mangrove, or forest. Mark them well, for you will see them only here and in Java and Sumatra.

# Common Iora
*Aegithina tiphia*
TOP RIGHT

Hear the long drawn-out mellow fluting whistle, increasing in pitch and ending abruptly on a lower note: *tweeeeeeeeeee-tyu*. It is but one expression of the Common Iora's varied vocabulary of fluting notes and bubbling churrs, betraying the presence of this cheery, if diffident, greeny-yellow bird prowling about the thick foliage of fruiting trees.

Every garden on Bali has its resident pair, and the mangrove at the end of Sanur beach is a favourite haunt — this area affords good views of many birds and the chance to hear their full repertoire of whistles and churrs.

# Ashy Tailorbird
*Orthotomus sepium*
BOTTOM LEFT

Also working their way through the shrubs, now furtively creeping, then boldly darting, in search of grubs, are the tiny grey Ashy Tailorbirds, distinguished by their elegant posture, jauntily cocked tails, and rufous faces. Noisy things they are too, emitting a sound like the anguished buzz of a fly trapped in a spider's web, and a regular loudly trilled: *tree-yip*. Often they enter the house and help themselves to kapok stuffing from my cushions, with which to add some comfort to miraculously stitched leafy nests in the hedgerow.

# Oriental White-eye
*Zosterops palpebrosus*
TOP LEFT

This is another little all greenish-yellow bird with a bold white eye-ring that regularly visits gardens from sea to mid-levels, usually in small rowdy bands sounding much like a battery of day-old chicks. At higher elevations the Oriental is replaced by the Mountain White-eye which has the belly whitish grey, though there is evidently some overlap, for I see both kinds, often together, in the Ubud area.

# Pied Fantail
*Rhipidura javanica*
BOTTOM RIGHT

Another familiar garden bird, forever on the move, posturing and pirouetting, flirting and fanning its tail. Despite its pretty performance, it has an aggressive side to its nature, and I have often watched it chasing other birds, including its own kind, venting an angry chatter like that of a swallow disturbed on the nest.

The Pied Fantail also has a pretty lilting song: *tum-titty tum-titty tweet tweet*.

# Scarlet-headed Flowerpecker

*Dicaeum trochileum*
BOTTOM RIGHT

One of the most brilliant little birds, though it is not always easy to see on account of its rapid flight and tendency to stick to the tree-tops, is the Scarlet-headed Flower-pecker, the males with bright vermilion scarlet heads, breasts and backs, and the females more mousy, having the rump red only. Invading every garden, a sharp, brittle clicking note uttered in flight always heralds their appearance: alighted males emit a thin and very high-pitched: *wheety wheety wheet.* They delight in the berries of *loranthus,* a type of tropical mistletoe whose growth they help to propagate, and may best be seen feeding in fig and other fruiting trees festooned with the stuff. The nest is a pale, downy, purse-shaped affair, suspended from the leafy twig of tamarind or acacia. Inhabiting South East Borneo, Java and Lombok (plus one or two other small islands) only, outside Bali, the birds are familiarly known here as *tabia tabia,* the local name for chilli.

# Olive-backed Sunbird

*Nectarina jugularis*
TOP

It is inconceivable that any visitor to Bali could fail to notice this radiant aerial wizard flitting from flower to flower, hovering before a selected bloom as a New World hummingbird, and inserting its long pointed bill and longer tubular tongue to extract the nectar on which it mainly feeds. But sometimes a short cut is taken, and the bill is used to stab the base of the staminal column. In order to appreciate fully the appearance of the male — the female lacks the metallic-coloured plumes — it is necessary to see him briefly quiescent on limb of

hibiscus, twittering and tweeping, and preening, in soft late or early light. Then, and only then, are his true colours revealed, for the forehead, throat and upper breast which at first seemed black, now assume a glittering lustre, irradiating half the colours of the spectrum, from greenish-blue, to indigo and violet; surmounting a bright mustard waistcoat, adorned on either side with gaudy pectoral tufts providing vivid orange patches for pockets.

# Brown-throated Sunbird

*Anthreptes malacensis*
LEFT

Widely distributed and quite common, Bali's only other sunbird, the Brown-throated, is more often heard than seen. It is greatly attracted to coconut palms, where it spends much of its time taking the insects which congregate on the branchy flowering sprays. In good light the male exhibits indescribably lovely hues of metallic green, purple, violet and maroon. A bulkier bird than the Olive-backed, the plainer dusky olive-green and yellow females are easily separated from the latter by the absence of white tail-tips. The call is an endlessly repeated plaintive: *wheet-tiu wheet-tiu wheet-tiu* — putting me in mind of the British chiff-chaff. To both our sunbirds the Balinese give the name *kepetjit.*

OLIVE-BACKED SUNBIRD

JARUIS.

# Magpie Robin

*Copsychus saularis*

Taking breakfast on the verandah the morning after arrival, the hotel guest is sure to be greeted by this conspicuously pied and sprightly vocal visitor to all the gardens of the East. The long-tailed, thrush-like, black and white bird glides low between the trees and alights on the lawn. Now it fans and cocks its tail, gazing expectantly about, before pottering off in search of worms and other succulent morsels in the top-soil and tussocks of grass.

The male Magpie Robins are strongly territorial, proclaiming their title by constant outbursts of clear whistled song and harsh scolding notes, effectively telling intruders to buzz off. One of the more familiar stock phrases may be articulated as: "woe betide you, won't abide you" — whence the popular onomatopoeic name, *betjitje*.

It is worth noting that our Bali birds have only the wing patches and outer tail feathers white, whereas their northern cousins in Java have white bellies, but intermediate forms occur. In the females the black plumage is replaced by grey.

# Black-naped Oriole

*Oriolus chinensis*

TOP

"Toodle-oo-oo, toodle-oo-oo" comes the valediction from on high. The melodious liquid call draws our attention to the large golden yellow bird with a black head-band, sitting in the tree-top. The sound is full and mellow and carries far, at times compressed into a loud squeal or slurred legato note, reminding one of a street urchin's wolf-whistle. But, like the Magpie Robin, this oriole may change its tune abruptly, at one moment fluting eloquently, at the next cursing like a trooper.

Best observed in undulating flight between the crowns of lofty trees, usually in small parties or pairs, the females are duller with the mantle more olive, and young birds are whitish under, streaked with black. The local name is *tjilalongan*.

# Spotted Dove

*Streptopelia chinensis*

BOTTOM

Widespread and common in cultivated land where it is often seen in large flocks roosting in trees or feeding on the ground, this turtle-dove, brownish above and a lovely soft vinous shade below, is distinguished by its black and white chequered half-collar and broad white tips of the outer tail feathers, clearly visible when the tail is fanned on alighting. Introduced to Bali and many other places, the Spotted Dove is a popular cage-bird, valued as much for its soothing voice as for the medicinal quality of its droppings which are said to be an effective cosmetic, and this may be equally true of all doves and pigeons. I prefer to forgo the use of skin-conditioners and to see and hear the birds displaying and calling in the wild: *coo-coo-krooo-coo* or *coo-coo-kroo*, without the final *coo*. The Balinese name is *kukur*.

## Fulvous-breasted Woodpecker
*Picoides macei*
TOP

The woodpecker family is fairly well represented in Bali and will be of particular interest to our Australian visitors, since it is altogether absent from the Australian region. Of the six kinds of woodpecker recorded from Bali only one can be considered common and may be found, usually in pairs, wherever there are trees, especially coconut palms in which the nest-hole is drilled. You will always hear them first: either the sound of bill drumming against wood, like a loud rattling human snore, or a series of queer squeaky peeps, similar to the noise produced by a deflating balloon or Christmas tickler. The sexes are alike, except that the male has the forecrown bright red; in the female the whole of the crown is black. All the woodpeckers are known locally as *belatuk*.

## Pied Bushchat
*Saxicola caprata*
BOTTOM

Much like a miniature Magpie Robin, every expanse of rice-field and other cultivation will yield at least one male bird hawking insects from a favoured perch, invariably accompanied by his mate of quite different appearance — all earthy brown with buffy rump — while immature birds are mottled and spotted with various shades of brown. Characteristic chat traits include the habit of shivering the tail, and uttering a few harsh grating *tchak* notes besides the song, which consists of several brisk musical phrases delivered in a pleasing, breezy manner.

# Coppersmith Barbet

*Megalaima haemacephala*
BOTTOM

Interestingly situated on the very edge of the Indomalayan region, Bali is also the last outpost of the barbet family, related to the woodpeckers, of which we have four species, all basically bottle-green jungle birds with vividly coloured facial patterns. The Coppersmith is more widely distributed than the others, often visiting gardens and drawing attention by its incessant ringing *toonk toonk toonk* call, delivered at the rate of anything from 80 to 200 *toonks* per minute. I hear one now distantly dinning, as the beating of a hammer on an endless metal strip. The race inhabiting Java and Bali, incidentally, has far more crimson about the head and breast than birds elsewhere in South East Asia.

# Island Turtle-Dove

*Streptopelia bitorquata*
TOP

Found on islands from Java to Timor, the Island (also known as Javanese) Turtle-Dove is a bird of the lowlands, particularly common on the coast. Compared to the Spotted Dove, it is greyer above, more ruddy vinous below, and the hind half-collar consists of a black band bordered above with pale grey. You will see both kinds lined up for inspection on electric cables, soon telling apart their cooing calls: a more guttural unaccented *kroo-kroo-kroo-kroo* in this case, not unlike the vibrating bow-strings of the giant *banjar* kites.

# Plaintive Cuckoo

*Cuculus merulinus*

The call is so familiar, yet the bird so seldom seen. It spends hours on end sitting concealed high in a shady tree, pouring forth its melancholy song, consisting of four mournful notes in cadence, followed by a sequence of fast tumbling notes, from which the bird derives its Balinese name: *ngkik-ngkik-ngkir*. Another call frequently heard may be expressed in the words: "tea-for-three," repeated usually three times on an ascending scale. Like many other cuckoos, the Plaintive is parasitic, laying its eggs singly in a succession of nests. Those of Ashy Tailorbird and Zitting Cisticola are recorded from Java, and doubtless it is the same unwitting hosts that are put upon here.

Many are the myths surrounding this cuckoo in different lands. In Bali they say that *kedis kedasih*, the sad bird, brings forth her young alive upon the sea, and in so doing expires; and being aware of her unhappy fate she spends her life lamenting it. In reality though, only the male birds make the distinctive calls, and as far as I can tell the females do not utter a sound:

How happy are the cuckoo's lives
Since they are blessed with silent wives.

# Greater Coucal
*Centropus sinensis*
TOP

# Lesser Coucal
*Centropus bengalensis*
BOTTOM

These two closely related species are lumped together, having a somewhat similar appearance, and both being equally common, the Lesser probably the more abundant. They differ markedly in size, though this is not a useful feature in the field unless the two are seen together (which they often are), and the Lesser is paler and duller, lacking the bright chestnut and glossy black plumage of the Greater.

Often referred to as crow-pheasants, they could be mistaken for gamebirds, but they give off a rather disagreeable smell and I am told that the taste is foul; though this does not deter the Chinese from pickling them in rice spirit for unspecified medicinal purposes. It would be most unsporting to shoot them in any case, for they are extremely clumsy fliers, spending most of their time on the ground, sometimes clambering up low trees in order to gain height and launch themselves into the air.

Unlike their disreputable relatives, these ground-dwelling cuckoos do build their own nests — great balls of grass and tangled vine, well-hidden in dense undergrowth or the tall *alang-alang* covering the slopes of river valleys. And here of a late afternoon will you be sure to see the big birds, sitting exposed and venting their familiar booting calls — in the case of the Greater Crow-Pheasant, a hollow *hoop-hoop-hoop-hoop*; and the Lesser, a staccato *kok-kok-kok-kok-kok*, followed by an eruptive *kookokgan kookokgan* — whence the name *kookokgan*, generally applied by the Balinese.

# Long-tailed Shrike
*Lanius schach*
TOP LEFT

# Brown Shrike
*Lanius cristatus*
TOP RIGHT

The scourge of all small creatures including birds, *tjeride* watches and waits, cruel hooked beak and wickedly sharp claws ready to pounce, then seize and transport its captured prey back to the larder, impaling it on a thorn for leisurely consumption at a future sitting. You may spy this miniature bird of prey, perched prominently in any expanse of open ground, but your unwanted presence is certain to provoke an outburst of muttered curses and harsh scolding notes, accompanied by vigorous tail wagging.

Ranging widely through India and South East Asia in various forms as far south as Timor, our bird is beautifully marked with the crown soft grey, above the usual black mask, extending on to the mantle; and underparts vary from pale salmon to rich rufous pink. The shrike tribe is not represented in Australia.

Dry areas are the favoured haunt and the open grassland around Nusa Dua is full of shrikes, many of them foraging after the manner of thrushes and starlings on the ground. Here may you also spot the smaller migratory Brown Shrike, from October to April, all brown and white, with distinctive white eyebrow above the black mask.

# White-breasted Waterhen
*Amaurornis phoenicurus*
BOTTOM

Secretive like other members of the crake family, this common bird does, however, often come out in the open and may be seen plashing through the paddyfields in the early stages of growth, pottering about the garden, or dashing across the road. And I have on occasion even caught it in the act of thieving scraps from the kitchen.

For all its ungainliness, yet it has a certain elegance of gait, proceeding deliberately on its splay-footed way, at each step, jerking and twisting the head, tail jauntily cocked to reveal full cinnamon flanks and under tail coverts. All at once it freezes, inclines the neck to either side, and with shoulders hunched and head bowed, scuttles off and melts into the contiguous green growth.

Were we unable to get a glimpse of the White-breasted Waterhen, we could not fail to hear it. The noise it makes is stupendous, and the name *tjekruwak* is an apt rendition of it. All day long, and sometimes well into the night, beginning with a series of squeals and grunts and gurgles like dirty bath-water escaping down a blocked drain, we hear *tjekru*, followed by *wak-uwak-uwak*, and again *tjekru-wak-uwak-uwak-uwak*, and so on, a thousand times or more. It is not so much a chiming in, but rather a cranking up. But by now, one is fast asleep.

BROWN SHRIKE, TOP, AND LONG-TAILED SHRIKE

# White-bellied Swiftlet
*Collocalia esculenta*
BOTTOM LEFT, FLYING

A mass of small black birds mills round the spreading crown of a banyan tree, flying swiftly on slender sickle-shaped swept back wings, alternating glides with a few rapid bat-like wingbeats. They are swiftlets, and it is a fairly safe bet that we have here the White-bellied Swiftlets which are widespread and abundant in Bali. Viewing them closely in a good light, the white bellies are plainly seen, and as they swoop low over the fields the black backs appear to be greenish glossed. There are other swiftlets, including all blackish Edible-nest Swiftlets, whose nests are harvested from coastal caves and used in the preparation of bird's-nest soup. Identification of the various species poses a problem even for experts, and I know eminent ornithologists who throw up their hands in horror whenever the topic is mentioned.

In the late afternoon the White-bellied Swiftlets skim over the rice-fields, feeding on the swarms of small insects ascending into the air, and uttering low guttural chitters; and so they continue until nightfall, the last of the day's little birds on the wing, fluttering in the fading light and mingling with the bats which they now so nearly resemble.

The swifts and swiftlets are often confused with swallows, and it takes time for the beginner — I too was once a beginner! — to sort them out in the field. For a start, their flight is much swifter, sustained and more direct. With swallows the wingbeat is shallower, and they glide with wings partially folded, which swifts never do. Another difference lies in the fact that the swallows are perching birds, whereas swifts with their small feet and curved claws can only cling to the vertical surfaces of cliffs and caves where they roost and build their nests. So we may be certain that those swallow-like birds sitting on the fence and overhead wires are emphatically swallows — not swifts.

# Barn Swallow
*Hirundo rustica*
CENTRE RIGHT, PERCHED;
TOP CENTRE, FLYING

# Pacific Swallow
*Hirundo tahitica*
BOTTOM, PERCHED; CENTRE, FLYING

# Red-rumped Swallow
*Hirundo striolata*
TOP RIGHT, FLYING

The swallows are really too well known to warrant detailed description, and who in northern climes is not familiar with these welcome harbingers of warmer weather? And who, on first witnessing their annual congregation, has not been touched by melancholy at the thought of fleeting summer? But we are untroubled by such thoughts in Bali, where the cosmopolitan Barn Swallow visits from September to April, and the Pacific Swallow, ranging from India to Polynesia, is resident. To distinguish them, the former generally has much longer tail streamers, bright white unders, and a black band across the breast, while the latter is greyer under and lacks the black band between chestnut throat and pale breast. A third kind visits and may even breed here: the Red-rumped Swallow, larger and more heavily built, with pink rump and dark streaking of the white underparts. Sweeping overhead, it emits a loud rattle strongly reminiscent of maracas. The swallows collectively are known by the local name *sesapi*, and the swiftlets, *griti*.

JARVIS.

# Singing Bushlark
*Mirafra javanica*
BOTTOM

Ranging all the way from Africa to Australia, the Singing Bushlark is the one member of the lark family to be found in Indonesia. A very curious feature of this little brownish streaked bird with a short tufty crest is its chameleon-like quality of adapting plumage hue to match that of the soil wherever it is resident. So it may vary from pale sandy to rufous and a dark earthy brown: notice also the typical lark facial character of palish eyebrow extending in a semi-circle behind and below the eye.

It is a glorious songster, capable of holding its own with any skylark. Visiting the rice-fields in considerable numbers when seasonally dry and at harvesting, you may hear its infinitely varied, sweet and high-pitched warbling song constantly uttered in fluttery flight and even while the birds are on the ground.

# Zitting Cisticola
*Cisticola juncidis*
CENTRE, ABOVE

# Golden-headed Cisticola
*Cisticola exilis*
CENTRE, BELOW

Both of these tiny warblers, known also more euphoniously as fantail-warblers, frequent the rice-fields, and there is a mad scramble for nesting sites as the shoots attain height. The nests are slight domed structures, artfully concealed in the heart of a selected clump by binding the stems together with spiders' gossamer.

Virtually impossible to tell apart in winter plumage, but observed when breeding, the Golden-headed males wear bright buffy-golden caps, while those of the Zitting are brown, strongly streaked black like the rest of the upperparts. So can they be recognised in different areas, according to the crop cycle, nearly all the year round. By their songs are they more readily identifiable: the Zitting tumbles through the air with a queer clicking *zit zit zit*, while the Golden-headed sits and emits a scratchy, insect-like buzz followed by a clear belling note. *Tjetjetrung*, the Balinese name, describes it well.

# Java Sparrow
*Padda oryzivora*
TOP

A prosaic name for such an outstandingly decorative bird. Unhappily popular pets, I think everyone must know these immaculate grey birds, with wine-coloured bellies, black and white heads, and massive bright pink bills. But by no means everyone will have seen them in the wild, and fewer people still may be aware that they are indigenous only in Java and Bali. Now, however, they have been introduced to many other places. Small parties of them frequently visit the garden, and how delightful it is to see these brilliant birds flying freely. The Java Sparrow is known generally throughout Indonesia by the name *jelatik*.

*SINGING BUSHLARK*

# Javan Munia
*Lonchura leucogastroides*
CENTRE

The munias are finch-like birds of the Old World, closely related to the sparrows and weavers. But unlike the finches, which construct cup-shaped nests, munias build elaborate covered affairs, usually colonially in some numbers, seeming to favour ornamental palms planted near restaurants and pools in hotel gardens. More colourfully known as manikins, these diminutive birds are equipped with huge conical bills designed to crush grain and grass seeds.

Distinctively dressed in dark chocolate bibs bordering white bellies, the Javan Munias are found only in Java, Bali, Lombok and the very south of Sumatra, though they have recently been introduced into Singapore. A serious crop pest, the farmers' lads are encouraged to destroy the nests wherever they may be located, which is sad but understandable.

JAPANESE SPARROWHAWK WITH A MUNIA

# Scaly-breasted Munia
*Lonchura punctulata*
BOTTOM

An unfortunate if diagnostic name adopted in most current publications, as if it were prone to a periodic reptilian sloughing, or suffering from dandruff, the Scaly-breasted is known from the older literature as the Spice Finch or Nutmeg Manikin. Its geographic range extends all the way from India, through South East Asia, to Timor and Tanimbar, and it has been introduced into Australia, where it is well established in Queensland and New South Wales.

The most abundant munia in Bali, it occurs widely in grassland and cultivation, forest, scrub, and mangrove. Unlike the Javan Munia which seems to be a rice ravager, and although observed in mixed flocks feeding openly in the paddy, the Scaly-breasted feeds mainly on the small seeds of various wild grasses that spring up irregularly with the rice crop, performing a valuable service and doing no damage to the paddy.

# Chestnut Munia
*Lonchura malacca*
TOP

# White-headed Munia
*Lonchura maja*
CENTRE, ABOVE

It is indeed confusing that the subspecies of Chestnut Munia found in Java and Bali should have the head white, whereas elsewhere it is black. With the White-headed Munia, the white extends to the chin and throat, while the former has these parts black. Both kinds invade the fields in swarms, the Chestnut mainly at mid-levels, the White-headed in coastal *sawah* near the mangrove in which they nest. The Balinese know them as *prit bondol*, the general name for all munias being *prit*, an approximation of their thin little chirping calls.

JARUPS.

# Cinnamon Bittern

*Ixobrychus cinnamomeus*
BOTTOM LEFT; TOP RIGHT, FLYING

Seldom seen in the open, the bitterns are shy and solitary denizens of the paddy in an advanced stage of growth, and fresh-water swamp, though the latter habitat has been all but eliminated in Bali, where the birds have become quite well adapted to mangrove. When danger threatens, they have the remarkable habit of stretching their bodies to the fullest extent and freezing, with the head and bill pointing skyward. And they will maintain this stance for as long as need be, their tawny streaked appearance blending perfectly with the surrounding vegetation so as to render them virtually invisible.

The Cinnamon Bitterns are quite plentiful whenever the rice is high and are easily flushed, flying low over the fields, when they cannot be mistaken for their uniform rich rufous appearance. In the late afternoon they emerge and stand on the bunds, immobile in the fading light, evenly spaced throughout the area and invariably at the same spot. Sometimes they roost in trees in the garden, and even on the roof.

The only utterance is a strange bubbling cadence, like the sound of a full bottle being poured till empty: *kok-kok-kok-kok-kok-kok...*, with neck extended horizontally, bill shut and throat throbbing. The delightful Balinese name is *kokokan maling*, the latter being the word for thief. By its secretive skulking habits and peculiar creeping gait, it certainly seems to be up to no good.

# Yellow Bittern

*Ixobrychus sinensis*
CENTRE

Still more elusive than the last species, and usually seen only when flushed, in flight the contrasting yellowish buff and black wing pattern is plainly visible. A winter visitor from the North, the Yellow Bittern is with us in some numbers during the early part of the year, mainly in southern coastal *sawah*, and I used to see many in marshy tracts bordering the central highland lakes, but practically all of that wonderful refuge for water-birds has recently been reclaimed for cultivation.

# Ruddy-breasted Crake

*Porzana fusca*
BOTTOM RIGHT

Another extremely secretive bird, yet it occurs far more commonly in the rice-fields than one might think, and the dog helps to put them up on occasion. Like the bitterns, they emerge from the crop cover late in the day and sit on the bunds preening, so as to afford fine views of the rich red chestnut head and breast, and delicately white-barred flanks and belly. In marshy patches of mangrove and overgrown fishponds betwen Sanur and Kuta, both Ruddy-breasted and White-browed Crakes may often be glimpsed, the latter light brown above, pale grey below, with two striking white stripes on the side of the head. More often heard is the noise they make: a succession of reedy piping notes, like a squeaky rubber-doll.

YELLOW BITTERN

# Purple Heron
*Ardea purpurea*
TOP; CENTRE RIGHT, FLYING

# Grey Heron
*Ardea cinerea*
TOP RIGHT, FLYING

Any large dark heron-like bird seen ponderously flapping along on broad rounded wings, neck tucked in an S shape and long legs trailing behind, is almost certain to be a Purple Heron. It is much larger than the Reef Egret which might be found in its dark phase and in the same neighbourhood. The general appearance is drab, but at close quarters the soft rufous shading of head and neck and darker chestnut underparts, and the blackish line running down the sides of the long kinked neck, are conspicuous. Very shy, at your least approach the great bird will take off, revealing the wing pattern, grey with darker tips.

There are always a few Purple Herons lined up on the mud-flats south of Sanur, and they are the first birds to take up their posts with the receding tide. And they are the last to return to their customary haunt in the adjacent mangrove and fish-ponds as the lagoon refills. Small parties of the similar-sized but much paler grey and white Grey Herons sometimes visit us during the northern winter.

# Little Heron
*Butorides striatus*
BOTTOM LEFT

The sinuous necks, pointed bills and long legs are typical of the herons; the legs designed for wading through shallow water in pursuit of fishes, eels and frogs, the neck curved or craned, and bill poised, ready to strike out on the instant and jab, grab and batter the catch senseless, before bolting it down in two or three great gulps.

The fishponds, mud and mangrove abound with the medium-sized dark grey Little Herons, standing stock-still on one leg, or creeping about furtively in a crouching attitude. The slow-motion searching, punctuated by sudden thrusts of destruction, is accompanied by constant tail-wagging. Truly a fascinating bird to watch.

With an exclamatory guttural *keeyo* or *kyok*, often repeated once or twice, the grey form erupts and flaps laboriously but a foot or two above the water's surface, trailing its yellowish legs, with oddly brilliant orange-soled feet; then plunges out of sight into the mangrove thicket, uttering a final croak.

# White-browed Crake
*Porzana cinerea*
BOTTOM RIGHT

Mentioned on the previous page as frequenting a similar habitat together with the Ruddy-breasted Crake, this marsh dweller still occurs in some numbers on the highland lakes, though greatly reduced owing to the encroachment of cultivation. Its long pale green legs and very long attenuated toes enable it to walk on floating vegetation. Also uttered, often by several birds in chorus, is a trill resembling that of the Little Grebe, or dabchick.

PURPLE HERON

jarvis.

JPARUIS

# Javan Pond-Heron
*Ardeola speciosa*
BOTTOM

From mountain lakes to the coast, it is an extremely common bird in every watery habitat, and large flocks are found feeding in the rice-fields, as well as in mangrove and on the mud-flats south of Sanur. In behaviour and build a bit like a bittern, but much more conspicuous either in motion or in a frozen attitude, head poking up above the sprouting rice. The breeding plumage of buffy head, cinnamon breast and black mantle is most distinctive, though non-breeding and immature birds, streaked and spotted white on buffy-grey, are not so easy to see. All remarkably erupt into white when they are put to flight.

The Javan Pond-Herons roost and nest communally with the three types of white egret discussed below, in the village of Petulu situated a couple of miles north of Ubud, and there should you repair to view them and all the thousands of other startling white birds, late in the afternoon when they return in formation from afar.

# Cattle Egret
*Bubulcus ibis*
TOP RIGHT

# Little Egret
*Egretta garzetta*
CENTRE RIGHT

# Short-billed Egret
*Egretta intermedia*
TOP CENTRE

The egrets are difficult to identify at the best of times, and to the untrained eye they all look alike: largish, long-necked, elegant and white. All three species often forage together in the fields, affording ready comparison, but a visit to the heronry at Petulu will provide an unrivalled opportunity to study them at close quarters and differentiate between them.

In their nuptial plumage more easily separable, the Cattle Egrets are mottled ginger, with bill and legs yellow to reddish; the more unsociable and aloof Little Egrets develop two long pendant head plumes, having bill and legs always black; and the Short-billed or Plumed Egrets, which tend to consort with the Cattle Egrets both when feeding and nesting, display billowing breast and back plumes, black legs, and bills varying from all yellow to black.

By behavioural traits and general appearance and posture — which birders term "jizz" — will the patient observer learn to tell them apart. The Short-billed tend to plod and have the habit of remaining motionless with necks held straight at an angle of 60 degrees or 10 o'clock, whilst the Little Egrets are seldom still for a moment, dashing hither and thither in pursuit of their quarry. And their bills are clearly longer and finer, exceeding the length of the head by about a third. To the Balinese all the white egrets are known as *kokokan*; the Javan Pond-Heron is called *blekok*.

FROM TOP: CATTLE EGRET, LITTLE EGRET (BREEDING PLUMAGE SHOWN AT LEFT), AND SHORT-BILLED EGRET

JARVIS.

# Crested Serpent-Eagle
*Spilornis cheela*
BOTTOM

Of the 19 different kinds of hawk on the Bali list, probably not more than eight are actually resident, while the status of but four is substantiated by direct breeding evidence, including the three described here and the Moluccan or Spotted Kestrel. A few years ago I would have said that the Serpent-Eagle was a relatively common bird, certainly the most frequently observed of all its tribe. Sadly I seldom see one now. Like all the raptorial birds they are miserably persecuted.

As one might assume from its name, the Serpent Eagle feeds mainly on snakes, though I am told it is not averse to the odd chicken, which if true could hardly be said to enhance its chances of survival. The Balinese name, *kekulik*, increasingly accented on the second and third syllables, reproduces perfectly the sound of its piercing scream which first attracts one's attention to the majestic presence wheeling overhead. Note the prominent white band on the trailing edge of the wing, and broad-banded tail.

I have forgotten to mention this eagle's peculiar habit, on alighting, of vigorously switching its tail in the manner of an agitated cat, which lends it a demeaning and comic aspect.

# Brahminy Kite
*Haliastur indus*
CENTRE

Formerly quite common on the coast, the Brahminy Kite is easily recognised by its striking chestnut plumage and contrasting white head and breast. In India the bird is considered sacred, though it cannot be said to be held in such reverence here, where it has an unenviable reputation as a stealer of chickens. Unfortunately, the dispossessed smallholder is unlikely to pause and reflect on the good it does as a general scavenger and preserver of the balance of life. I knew of a nesting site (since destroyed) in the last tall stand of mangrove at Pesanggaran, and here I have observed the perched adult birds uttering a curious cat-like call, neck held directly up and beak wide open. Another more strident note vented in flight has been blithely likened to "the squeal of an unoiled door hinge." Young birds lack the ruddy hue, being duller and mottled and noticeably patchy under the wings.

# White-bellied Sea Eagle
*Haliaeetus leucogaster*
TOP

A much larger and heavier bird than the last, equally distinctive in its piebald appearance and also frequenting our coastal waters, *Haliaeetus leucogaster* (Greek *hals*, the sea, *aetos*, an eagle, *leukos*, white, and *gaster*, the belly) is one of the most magnificent eagles to be seen anywhere in the world.

But even eagles are easily slaughtered by clowns with air-guns. Of course they should be protected by law, for it is clearly criminal to shoot such splendid creatures that are completely harmless to man.

Sailing off the eastern tip of Nusa Penida, I spied a pair of sea eagles standing sentinel by their nest, a huge scraggly pile of twigs strung in the topmost boughs of a tree, which leaned out dizzily from a sheer cliff over the ocean depths. There were two unfledged young in the nest, and I observed the parents in turn swooping down and returning to feed them, their loud clangorous calls reminding me very much of the honking of domestic geese. A mile or more further down the coast, another pair perched on their nest atop a tall stack, dashed by mountainous waves. And I reflected that we are indeed fortunate that in this over-populated land such remote sites yet exist to ensure the survival of the species, as they continually renew their nests and raise successive generations.

# Common Sandpiper
*Actitis hypoleucos*
BOTTOM

Seen in small parties on the beach, and inland usually singly or in pairs, seldom in the fields but more commonly teetering on rocks and pottering along the banks of fast-flowing streams. Who cannot be familiar with this small dull brown and white wader, as it flits in short bursts on vibrating wings, barred brightly white, uttering a plaintive, piping *twee-wee-wee?* Its constant rhythmic bobbing motion puts one in mind of a child's toy activated by a clockwork spring.

# Wood Sandpiper
*Tringa glareola*
TOP; FLYING

This and the Common Sandpiper are the two wading birds or shore-birds which visit Bali on passage and spread inland, not being restricted to the coast. Over-wintering here, many remain with us until around April and are absent only during the following two or three months.

I have seen thousands of Wood Sandpipers flocking in flooded *sawah* near Sanur and Kuta in April, and again in August, and fair numbers are always visible in the fields about Ubud throughout the period of their stay. The shrill, sibilant cry: *sif-sif-sif-sif*, uttered in flight and in chorus, is one of the most memorable sounds of the open countryside.

# White-breasted Woodswallow

*Artamus leucorhynchus*

TOP

A common bird of open country and woodland, it is distributed throughout Indonesia to Fiji and Australia, where woodswallows — a family quite unrelated to the true swallows — are well represented by a number of different species.

Dapper, compact, dark slaty-grey birds, with striking white rumps, breasts and bellies, they have a belligerent disposition; and I speak from personal experience, having had them hurtle at me, furiously squawking, possibly because I ventured too close to their nesting site. Often seen in groups huddled side by side, venting disagreeable creaking notes and constantly wagging their tails, once airborne they fly wonderfully well, soaring above the fields and gliding great distances on their long and pointed triangular wings.

# Philippine Glossy Starling

*Aplonis panayensis*

BOTTOM

The dark shining bottle-green bird with flaming blood-red eyes is unmistakable, though we may be less concerned with its appearance than its activity, as it helps itself to another liberal portion of ripe papaya in our garden. Occurring widely in lowland cultivation, these starlings nest in tree-holes, and immature birds differ in having whitish underparts heavily streaked with black. The closely related Short-tailed Glossy Starling, which probably visits us from the East, may be met with at higher elevations, and is very difficult to tell apart. It is more uniformly dark green with a purplish collar, while the other gives off strong blue reflections, notably on the breast.

# Asian Pied Starling

*Sturnus contra*
BOTTOM

Distributed from India to Bali, demonstrating yet again this island's peculiar and ornithologically intriguing situation at the frontier of the Indomalayan region. A bird of drier lowland areas and like others of its tribe a gregarious ground-feeder, I have seen it in large numbers in seasonally dry *sawah* in the south, and in the arid limestone zone (TheBukit) between Jimbaran and Ulu Watu. They used to nest in the Ubud area, constructing large, untidy platforms of straw and sticks in thorny coral-bean trees bordering irrigation canals, where they were well protected by a combination of toxic spines and vicious soldier ants from grasping hands, but not from the appalling air gun.

# Black-winged Starling

*Sturnus melanopterus*
CENTRE

A bird of far more restricted range, recorded only from Java, Bali, and Lombok, though its presence in the last island, whether as visitor or resident, requires confirmation by anyone interested enough to go there and look for it. As a footnote I might add that there exists on St. John's Island, just south of Singapore, a flock of 40-odd birds; so the species may have been introduced there, or perhaps it spontaneously expanded its range.

Typical Bali birds have white head and unders, grey mantle, and black flight feathers. However, to complicate matters, two further races visit us from Java, in which the grey is partly or wholly replaced by white; so be on your guard when identifying this species of such variable plumage. It is easy to mistake all white birds, with black showing only on wings and tail, for Bali's rare endemic Rothschild's Myna, or Bali Starling. Both species inhabit the same locality, the Black-winged Starling in very much larger numbers, and it requires no great stretch of the imagination to transform the obvious one into the elusive other. To disinterested locals they are both *jalak putih*.

Nesting in tree-holes and widely distributed, I have observed flocks of these starlings at every elevation, from the grounds of Nusa Dua's resort hotels to the greens and fairways of the Bali Handara golf-club.

# White-vented Myna

*Acridotheres javanicus*
TOP

Another bird of very variable plumage and doubtful classification, some authors treating it as a race of the Indomalayan Jungle Myna (*Acridotheres fuscus*). Confined mostly to coastal dry cultivation and grassland, where it is often seen foraging in association with cattle, similarly to Cattle Egrets and Black Drongos, these mynas vary from nearly all black to greyish-brown, presumed to be adult and immature plumage respectively. As a rule they show very little white on the vent and have short crests.

BLACK-WINGED STARLING

# Bali Starling
*Leucopsar rothschildi*

When I first came to Bali some twenty years
ago, I must admit that I had never heard of
the island's celebrated one true endemic,
the Bali Starling, which points principally
to a lack of familiarity with the specifics of
our avifauna, if not absence of interest in
birds in general. But then I do not suppose
that anyone, apart from a select company of
ornithologists, could claim at that time to
have more than a nodding acquaintance-
ship with this rare and reclusive creature.

The dismal, ironic truth of the matter is
that only since its inclusion in 1967 in the
Red Data Book of endangered species has
the Bali Starling become the desired object
of unscrupulous collectors, and consequently
doomed in its wild, unfettered state to
extinction. Certainly there are other con-
tributory factors to *Leucopsar's* (white starling's) extirpation, including the encroachment of "civilisation" and destruction of
habitat. Climatic quirks too play their fate-
ful part. And one may in truth contend that
climate, in the end, is the debt of Nature to
all life on earth.

Starlings are birds of dry ground. Seques-
tered in a naturally arid part of the Bali
Barat National Park, our darling Bali
Starling eschews water, save for its daily
intake of morning dew. I have heard a
remarkable and reliable account (which I
do not think has previously been published)
of a catastrophe that took place one day,
probably early in 1970. This date is postu-
lated in the light of its appalling aftermath
which I shall presently relate.

To provide food for the expanding local
population, a plot of savannah had been
burned and freshly turned in for crop planting; and on this swath descended an
immense flock of starlings of various kinds
to feed on the insects crawling in the
exposed soil. On a sudden there came a
cloud-burst and downpour of such fearful
intensity that the assembled host had no
hope of escaping. Both stunned and water-
logged, the wretched birds made ready pick-
ings by the sackful. And doubtless all would
have been dispatched and eaten, but for the
Bali Starlings whose value was too well
known, a fate far worse awaited. In that year
over 100 were shipped to America alone.
The unofficial figure must be greater. By
one fell stroke, perhaps as much as one half
of the entire population was wiped out, and
as this never amounted to more than "several hundreds," there is sadly little or no
hope of recovery. Less than 50 birds now
remain (October 1988) at liberty.

At least we may be confident that the
species will survive, for it is a fact that this
robust bird thrives in confinement, having a
total world captive population of nearly
1,000, and an elaborate scheme is afoot to
rehabilitate captive-bred birds, under the
auspices of the International Council for
Bird Preservation and the Indonesian
Nature Conservation Service.

One can only wish that the efforts of so
many dedicated people will be rewarded,
but the dilemma persists; for so long as our
*jalak putih* has a high price on its head (currently 300 or more U.S. dollars), and there
are scoundrels ready to pay it, poachers will
risk their necks to trap the birds.

At the same time it is wrong to be wholly
pessimistic. And it may well turn out that
depletion of pure native stock will deaden
the market attraction.

JARVIS.

# Blue-tailed Bee-eater

*Merops philippinus*
CENTRE

Exotic and gorgeously arrayed in subtle hues of blue and green, with long central tail feathers and coppery underwings, the Bee-eaters are essentially an African family, of which one species, the Blue-tailed, ranges from Africa, through India and South East Asia to New Guinea. Supremely graceful flyers, and often seen in flocks of up to several hundreds, their diet consists mainly of bees and wasps which they hawk on the wing from exposed perches such as leafless trees and electric cables.

On many occasions have I watched one of these birds swoop down and seize a large wasp or hornet, and on returning to its perch, batter and scrape the insect to remove the sting, before swallowing it or presenting it to mate or offspring.

A lively, liquid double note: *quillup quillup* always announces their coming in scattered parties, now soaring overhead in gently undulating flight, then sweeping low over the roof and across the intervening fields, to disappear beyond a distant ridge.

# Bay-headed Bee-eater

*Merops leschenaulti*
BOTTOM

Bali is the absolute limit of the range of this species, which extends from India, through South East Asia to Java, bypassing other parts of Indonesia. Here it seems to be confined to the western part of the island, requiring a trip to the National Park, where it frequents the forest edge in smallish flocks. There is no mistaking the bright chestnut bay crown and mantle, the typical wavy flight and lilting *lillip lillip* as they perform their manoeuvres, or hawk bees and other buzzing things from a favoured perch. Unlike the Blue-tailed, the central tail feathers of this species are not elongated.

# Dollarbird

*Eurystomus orientalis*
TOP

Known from India to Australia as the Dollarbird on account of the small round silvery-blue wing-patches, visible only when the wings are opened, this glossy dark greenish-blue bird usually appears all black when seen at rest. Then, only the bright coral-red bill and feet are conspicuous. So often seen in silhouette, sitting atop a dead palm stump, it is a weird-looking bird, big-headed, neckless and legless, instantly recognisable from afar. And yet it is a most graceful and acrobatic flyer, adept at rolling (whence its other name, Broad-billed Roller), as it glides upward, and seemingly about to stall, swoops down and forward and continues in similar vein, much as do the doves in their courtship flight display.

Also restricted to the National Park, in fairly open scrub, it sits about for the greater part of the day, at intervals venting queer frog-like croaks, and occasionally sallying forth after insects. Only late in the day does it become more active, indulging in its rolling flight and uttering hoarse squawks. Young birds have the feet and bill black.

BLUE-TAILED BEE-EATER

# Black Drongo
*Dicrurus macrocercus*
BOTTOM

In fact all the drongos are black, excepting the Ashy, and glossy or spangled: large arboreal birds of the Old World tropics, which almost demand attention by their striking appearance and posture and "in-flight" antics, as well as by their extraordinary vocabulary of metallic chimes and rasps. The Black Drongo is one of four species found in Bali, at the very limit of its range, and the most readily observable, for, unlike the other drongos, it prefers dry open country. Two areas in which you will surely see them are the National Park and the Bukit, south of the airport. For some reason the very mention of the name "drongo" conjures up a mental association of the spanking white rump of a Balinese cow. And so may you see the bird incongruously mounted, as witness the delightful detail in Frank Jarvis's portrait.

# Ashy Drongo
*Dicrurus leucophaeus*
CENTRE

Though more of a montane forest dweller, the Ashy is locally not uncommon in relict vegetation, thick and lush in river valleys. I used to spend many pleasurable hours observing a group of six to eight of these drongos from my house in Ubud. Their spectacular hawking flight and frequent dashing sorties to ward off intrusive crows impressed no less than their constant outbursts of sound: *tjeep tjurr-tjurr tjeep tjurr-tjurr*, interspersed with shrike-like rasps and peculiar squeals, and rapid reels of seven or eight notes up and down the scale.

But where are they now? I fear they have been dispatched, for they attracted too much attention, providing sitting targets for the intrepid "sportsman," equipped with airgun and an endless supply of slugs.

Together with the following species, the Ashy Drongo has managed to cross the Wallace Line, but no further than Lombok.

# Hair-crested Drongo
*Dicrurus hottentottus*
TOP LEFT

Distributed in several races from India to Australia, where it is the sole representative of the drongos, this is the bird I had in mind when I wrote the story, King Crow and Pyedog (*The Haughty Toad & Other Tales from Bali*), whose plumes were "all glossy green and purple sheen" and whose voice was as "the rasp of a saw on a split bamboo." It is the noisiest drongo of the lot, having an incredible vocabulary of buzzes, rasps, and trills, dominated by a resonant *klung klok-klok*. Also the glossiest, the spangled crown, nape and breast giving it a sort of two-tone appearance. Note the curling tail-tips, milky eyes, and long straggly hair-like feathers sprouting from the base of the bill. Found only in the National Park and on Nusa Penida, where it is quite common.

# Greater Racket-tailed Drongo
*Dicrurus paradiseus*
TOP RIGHT, FLYING

Another Indomalayan species whose last outpost is Bali, I have seen it only in forest in the central highlands and National Park. The most spectacular drongo with outer tail plumes extending a foot or more in the form of naked quills vaned only at the tips, other authors have likened it in flight to a blackbird being chased by two bumble-bees. At a distance the shafts are all but invisible; however beware of confusing it with other drongos, for the birds have a habit of breaking them. Familiar utterances are a clear ringing *tiu-tweep*, and a not-so-pleasing *tjee-tjee-tjee-tjwah-tjwah*.

# Pacific Reef-Egret

*Egretta sacra*

The Reef-Egret is a scientific curiosity in that it is dimorphic, occurring in two distinct forms or colour phases: dark grey and white. In the dark phase, which greatly outnumbers the white, the bird is all dark slaty-grey, with a blackish bill and conspicuous greenish-yellow legs, which trail noticeably in flight. Occasionally, birds are seen with white on the chin and throat, and we have one record of such a bird in Bali.

White phase Reef-Egrets are pure white, with yellow bills and the same greenish-yellow legs, which should serve to distinguish them from the other white egrets if found together. Their habits also differ from other egrets, Reef-Egrets being solitary and much more heron-like, crouching stock-still while foraging, with brief darting forays into the water, sometimes becoming completely immersed. Confined to reefs and rocky shore-lines, though I have seen them on mud-flats, usually the reefs at Sanur and Nusa Dua will yield one or more, as will the rocks jutting out into the ocean below the cliff temple at Ulu Watu.

JHRUPS.

# Common Pipit
*Anthus novaeseelandiae*
BOTTOM

Ranging from Africa to Australia and New Zealand, the Common (also widely known as Richard's) Pipit is a bird of dry grassland near the coast, though it has been recorded from Kintamani. The best place to see it is on the grassy peninsulas at Nusa Dua, where it is quite plentiful either singly or in pairs, moving briskly through the tufts in search of insects. These slender little brownish streaked birds may easily be recognised by their habit of standing very erect on long pink legs, surveying the ground about them for the slightest movement. On your approach, they will scurry off or take wing with a sharp, nasal *tjwizt*, on a short undulating flight, much like the wagtails to which they are closely related.

# Striated Warbler
*Megalurus palustris*
TOP; FLYING

Patchily distributed from India through parts of South East Asia to Java and Bali, this warbler seems equally at home in highland open bush vegetation and dry coastal grassland, where it may often be observed together with the last species.

Rather large for a warbler and long-tailed, it has an ungainly laboured flight, fluttering down heavily into low cover with wings and tail spread like a miniature pheasant. Its husky dark-streaked look also puts me in mind of the European Corn Bunting, with which it has in common the habit of sitting on an exposed perch, venting its queer ventriloquial call: a series of "jug" notes ending with a strident screech that carries far and wide, and from which is derived the local (Sundanese) name *tjek-tjekkorek*.

JARVIS.

# Eurasian Tree-Sparrow

*Passer montanus*

This bold, familiar little bird is so well-known that it scarcely requires an introduction. The term Eurasian denotes the original range of the Tree-Sparrow, from Europe to Bali, but, in common with its first cousin the House Sparrow (*Passer domesticus*), it has been introduced into other lands, including America and Australia.

Here we have a good example of how the habits of one species may vary from place to place, for what holds true of the Tree-Sparrow in, say, England, is certainly not the case here. In those parts of their range where they must compete with their slightly larger and more powerful relative, they are shy birds of the field and countryside. Without such competition in our region, they adopt the habits of the House Sparrow, and are always to be found where there is human habitation, even at the heart of the busiest city.

Nesting in trees in Europe, the description "Tree-Sparrow" is something of a misnomer here, where it prefers to construct its untidy pile under the eaves of buildings, having an apparent predilection for the roofs of modern government office blocks.

JARVIS.

# Savannah Nightjar
*Caprimulgus affinis*

Also associated to some extent with buildings, particularly at dusk when fair numbers of nightjars emerge to hawk insects attracted to the electric lights. Indeed the best place to view these large and graceful flyers is in the beams of the powerful spotlights mounted on the roof of the Bali Beach Hotel; and one's attention is invariably drawn to their peculiar nasal screeching calls, constantly repeated and penetrating: *djweenk djweenk*.

Another curious trait of these birds is that they choose to remain on the ground throughout the day. Well camouflaged by their mottled plumage, one may almost tread on them before they soar into the air, venting always an indignant *djweenk* or two in circular flight, then alighting again on the ground a short distance away. In the sparse growth of coastal dunes do they roost and nest, building no fabric, but laying and

incubating their eggs on the bare earth. These are also protectively hued, as are the chicks which resemble nothing so much as brown blobs of mud, and are quite impossible to detect even when one is gazing directly at them. Should one venture near a nesting site, the adult birds make a noisy display and ferociously attack the intruder.

Owing to their unfortunate habit of sitting at night on roads, eyes reflecting red in the headlights of oncoming motorcars, a great number are squashed flat. Found only in the National Park, another species, the Large-tailed Nightjar, is best known for its distinctive, resonant *tjonk tjonk* call, and one may enjoy good sport in betting on the number of "tjonks" expressed.

# Golden-bellied Flyeater
*Gerygone sulphurea*

The thin, silvery thread of sound, slightly wheezy and slurred, betrays the presence of this will-o'-the-wisp of a bird. Even at the hottest part of the day, the elfin notes slide up and down the scale, lulling the listener into siesta. It is the song of our one representative of the Australian family of *Gerygone* (from the Greek, meaning an echo born of sound) or Fairy Warblers. The Aussies call it "Sleepy Dick."

In Bali it is confined to the coast, frequenting the tall casuarinas bordering beach gardens, with a strong preference for mangrove. The best place to see and hear this tiny bird is at the southern end of Sanur Beach. If you stand very still in the shade at the edge of the mangrove, it will come and pose for you, immaculately dressed — all olive brown above, pale primrose yellow below — uttering its sweet song.

As you look on entranced, you may on a sudden be startled from your reverie, as the fairy form darts from its perch on whirring wings, to snap up a fly within a few feet of you, and with an unexpectedly loud "snap!"

# Streaked Weaver

*Ploceus manyar*

The weaver-birds are a largely African family, well-known for their construction of elaborate nests in large colonies, often suspended from palm-fronds or strung low in the tops of reed-beds.

Vast swarms of these finch-like birds noisily descend on the paddy from around August to December, when the males are easily recognised by the golden crowns of their nuptial plumage. Strangely, they seem to be outnumbered by the rather dingy females and non-breeding birds, which may be distinguished by their yellowish neck-patches, in the ratio of about 1:50.

The males are polygamous, taking as many mates as they can cope with, given the physical limitation imposed by the fact of their having to construct separate nests for each of their wives. Thus must they be considered amongst the most industrious creatures in all of creation. Recently I had a colony of them nesting in the coral-bean trees of the front hedge. It was quite an extraordinary spectacle, and the antics of the males would provide material sufficient to fill an entire volume ("The Weavers' Tale!")…and the noise…!

The familiar domed and funnelled nests are built with incredible speed — even within a period of 24 hours — and I have observed that they may be destroyed by jealous males in a matter of seconds. The females are totally indolent, merely sitting and preening and watching, and occasionally condescending to inspect the building in progress to see if it is satisfactory to them.

JARVIS.

# Arctic Warbler
*Phylloscopus borealis*
CENTRE

The two little *Phylloscopus* (from Greek *phullon*, a leaf, and *skopos*, a watcher) warblers — this and the next species — seem to spend all their time moving through foliage in search of insects. Known also as Leaf-Warblers and Willow-Warblers in more northerly climes, fair numbers of Arctic Warblers visit us from their Siberian breeding grounds in winter, journeying even to Australia, which is truly remarkable for such a tiny bird.

I have seen them here only in January and February, and then only in the National Park at sea-level. The ones I observed were a dull greyish-brown above, with just a hint of olive, and pale buffy under, with a distinctive whitish eyebrow. They reminded me of the British visitant Willow Warbler.

Prowling about in low scrubby vegetation, they emit a constant dry chipping sound: *dzib dzib dzib....*

# Mountain Leaf-Warbler
*Phylloscopus trivirgatus*
BOTTOM

Amongst the most difficult of all birds to identify, even when held in the hand, most of these Leaf-Warblers are greenish or olive above and greyish-white to pale yellow below, with contrasting light eyebrows and one or more fairly prominent wing-bars. Fortunately, having only one resident species here, the problem does not arise: moreover the Mountain Leaf-Warbler is a much greener and yellower bird than the others, and quite unmarked except for the dark stripes through the eye and over the crown, which give it a very distinctive head pattern.

An extremely common bird but localised, being found only in mountain forest of the central region, it may be seen foraging usu-ally in small parties through the lower canopy of tall trees, accompanied by a constant outpouring of thin chittering notes. To avoid the discomfort of a cricked neck, I can recommend that one lie down under a selected tree, training the glasses upward, and presently one may be rewarded by the sight of these active little birds, sometimes in mixed flocks with other species such as flycatchers, white-eyes and minivets, working their way methodically through a patch of jungle.

# Sunda Flycatcher-Warbler
*Seicercus grammiceps*
TOP

Restricted in range to Bali, Java and Sumatra, this attractive little bird of the forest is locally quite common and found in the same area as (and often together with) the last species. Its ruddy chestnut crown, soft grey mantle, spanking white rump and underparts, together with bold yellowish wing-bars, once beheld in a sunlit glade, are unforgettable. So too is the oft-repeated song, a series of very high-pitched whistled notes, delivered with beak wide open. Best observed by the Bali Handara golf course or the Bedugul gardens.

SUNDA FLYCATCHER-WARBLER

# Horsfield's Babbler

*Trichastoma sepiarium*
BOTTOM

The Babblers are an Old World family of the Indomalayan tropical region. Generally rather poor flyers, they tend to stay near the ground in dense cover, seldom coming into the open. Notoriously tricky to watch, it is a frustrating experience to have a flock of these birds muttering and grumbling in the bushes a few feet away, yet to be unable to catch even a fleeting glimpse of one of them.

Of the three species of Babbler recorded from Bali and here illustrated, Horsfield's is the most widely distributed, and though it is very much a Jungle Babbler, a relict population is locally common in the thick vegetation on ravine walls. Its presence is betrayed by the shaking of plant stems on which it alights, and by its peculiar utterances: a series of hoarse *kwik* notes, interspersed with muffled bickerings and melodious whistled snatches.

There seems to be some plumage variation and more than one race may exist in Bali, but generally it is a nondescript, short-tailed, plump, mousy-brown bird; whitish

CHESTNUT-BACKED SCIMITAR-BABBLER

underneath, with buffy flanks and dark greyish head.

# Pearl-cheeked Babbler

*Stachyris melanothorax*
CENTRE

Found only in Java and Bali, it is a very common bird of mountain forest in the central area, though it has been recorded from river gorges near Ubud. Noisy, babbling parties forage through the undergrowth and are briefly glimpsed flitting across jungle trails, but I have actually seen them swarming up trees to a height of sixty or more feet.

Attracted by the constant churring call-notes, we may try to obtain a view of one: all ruddy brown above, with brighter chestnut wings and crown, and underparts more greyish-brown, excepting the pale buff throat and upper breast which are conspicuously marked with a thin black line on either side, and wider black gorget. Our Bali birds have the cheeks light russet rather than pearl grey.

# Chestnut-backed Scimitar-Babbler

*Pomatorhinus montanus*
TOP

So called on account of its long, curved bill, this Scimitar-Babbler ranges from Malaya throughout the Greater Sundas to Bali. Its penetrating varied voice is one of the characteristic sounds of the highland forests: notably a hollow, ringing *kwo-ku kwo-kui* and an eerie, rhythmical, twanging thrum, which has exactly the timbre of a jew's-harp.

More easily observed than the other skulking babblers as it climbs in small hunting parties up trees, probing crevices in the bark for insects, note particularly the combination of rich chestnut mantle and white underparts, and black head adorned with bright yellow bill and white eyebrow.

# Fulvous-chested Flycatcher

*Rhinomyias olivacea*
TOP LEFT

# Asian Brown Flycatcher

*Muscicapa latirostris*
BOTTOM RIGHT

The first of these is one of those "little brown jobs" so dear to serious ornithologists, and which keep amateurs like myself on their toes. Restricted in range to the Malay Peninsula, Greater Sundas and Bali, it is an uncommon bird of highland forest fringe, and it has been reported as not uncommon in riverine woodland habitat near Teluk Terima in the National Park.

A typical flycatcher hawking insects on the wing, it is not in the habit of perching openly like the following two species, but sits for the most part concealed in low bushes and undergrowth. All olive-brown above and white below, it is distinguished by the broad brown band across the breast. Young birds are mottled or streaked with only the chin and throat distinctly white. Note carefully the characteristic bristled gape, and constant flirting of the tail. Care must be taken not to confuse it with a winter visitor, the Asian Brown Flycatcher, which is paler with a greyish cast and more uniformly whitish underparts; as well as with the Mangrove Whistler, which has the crown dark grey, throat and breast all ashy-grey, and a less erect posture. All three species may be found together in North West Bali.

# Grey-headed Flycatcher

*Culicicapa ceylonensis*
TOP RIGHT

There is no possibility of mistaking this conspicuous, confiding little bird, for it may always be seen poised expectantly on a low-lying limb, frequently sallying forth after insects, and even dogging one's passage along a forest track. Ranging throughout South East Asia to Bali (and possibly to Lombok: it would be nice to have confirmation of this), it is confined here to the Central Highlands and soon encountered by the golf links or in the botanical gardens.

Bright olive-green above and yellow-bellied, it is the soft, slaty-grey of the whole head, nape and upper breast, which draws attention, lending the bird an odd, big-headed look. The sweet song is of three or four notes: *whee whee whee-eet.*

# Snowy-browed Flycatcher

*Ficedula hyperythra*
CENTRE

From the Himalayas to the Moluccas, the tiny Snowy-browed Flycatcher adds a delicate grace to the high jungle. Still more disarmingly intimate than the last species, it sits by the trail until one is almost upon it — soft slaty blue, rufous pink, and snowy-browed — then flits off to another post, seldom seeming to rise above eye-level. The females are a lovely shade of olive-brown, with rusty highlights on wing and forehead. The song is a soft wheeze of several notes, followed by a sibilant *see-saw.*

MANGROVE WHISTLER

# White-winged Triller
*Lalage sueurii*
CENTRE; BOTTOM

The three birds featured on this page are members of the family *Campephagidae* or caterpillar-eaters (Greek *kampe*, a caterpillar, and *phagein*, to eat), which comprises the minivets, trillers, and cuckoo-shrikes. Replacing the Asian Pied Triller in East Java — there is no direct evidence of any overlap — the White-winged ranges through the Lesser Sundas to New Guinea and Australia. Favouring drier coastal areas, it seems particularly partial to casuarina and mangrove, and is surprisingly common on Nusa Penida.

# Small Minivet
*Pericrocotus cinnamomeus*
TOP; CENTRE RIGHT

Occurring from India to Bali, these beautiful, sprightly little birds, of which the males are soft grey and flaming orange, and the females duskier with orange rumps, may be encountered in small parties, briskly moving through the foliage of low trees with a constant outpouring of shrill, squeaky call-notes. Found often in mangrove and drier, more open country, they are more readily met with than the larger Scarlet Minivets which are strictly inhabitants of the forest.

# Black-winged Flycatcher-Shrike
*Hemipus hirundinaceus*
LEFT

Another small pied bird which could be confused with the Little Pied Flycatcher and Pied Bushchat, and even with the somewhat larger White-winged Triller and Magpie Robin, all of which could be found in the same habitat. The glossy black of the male becomes brown in the female, and the white rumps of both are distinctive in flight. In behaviour much like a flycatcher hawking insects from an exposed perch, its only utterance (I have heard) is a sibilant five-note: *tzi-tzi-tzi-tzi-tzi*.

## Mountain Tailorbird
*Orthotomus cuculatus*
RIGHT

Not an easy bird to see, this species tends to skulk in dense thickets of mountain forest. Yet it is not a bit shy, for it has happened on several occasions that one or a pair have emerged from wayside growth to forage within a few feet of me, quite ignoring my presence. The song is unforgettable, a haunting, quavering four note whistle in many different keys, consisting of two notes on the same pitch, followed by a drawn-out tremulous note and a short one, either higher or lower in pitch.

## Black-naped Monarch
*Hypothymis azurea*
TOP

Azure-blue with a thin black gorget and odd-looking black bob or "skull-cap," the male Monarch Flycatcher is unmistakable — his wife is brownish above and lacks the black markings. Noisy and obvious, attention is also unnecessarily drawn to the strident, nasal *tzweenk-tzweenk.*

## Golden Whistler
*Pachycephala pectoralis*
BOTTOM

Known also as "thick-heads," the whistlers are a typically Australian family, only one, the Mangrove Whistler, occurring throughout the Indomalayan region to West Bali and Lombok. The Golden Whistler is one other that has crossed the Wallace Line, establishing itself in Bali and East Java, and here it may be seen and heard only in the Central Highlands. The males are strikingly marked with black head and white throat, surrounded by a black band, brilliant golden-yellow below and olive-green above; the females much duller and greyer. To be serenaded by the glorious Golden Whistler is a memorable experience — a medley of poignant, piercing whistled notes, often delivered as a sequence, increasing in pitch and intensity, and ending abruptly with a remarkable whiplash effect. Seek out the source in the botanical gardens, and succumb to the magic of it.

# Little Pied Flycatcher

*Ficedula westermanni*

The sprightly little black and white birds —
all black above, with a very pronounced
white eyebrow extending to the nape, white
bar on the wing, and all white below — are
the males. Their wives are quite different,
drably dressed in greyish-brown mantle, a
touch of russet on the tail, and grey and
white below. But since they are often seen
in pairs, together with their mottled offspring, it should not be difficult to identify
them both.

They remind me very much of the
British Pied Flycatcher in well-wooded hilly
country of Northern England and Wales,
and here you will find them only in the
Central Highland region. Quite common
and not at all shy, they frequent the forest
edge and may easily be observed in trees
near the Bali Handara golf course or in the
botanical gardens.

The voice is variously described, consisting of a thin, sweet musical cadence of up to
ten notes, as well as insect-like wheezy nois-
es and burbles.

Not to be confused with the similarly
pied Black-winged Flycatcher-Shrike which
may be found in the same area, but which is
somewhat larger and lacks the white wing
and crown stripes.

JARVIS.

# Brown Honeyeater
*Lichmera indistincta*

The presence of honeyeaters on Bali is extraordinary and the subject of some debate amongst scientists, indicating yet again the peculiarity of the island's location at the frontier between two faunal regions — the Australian and Indomalayan (or Oriental as defined by Wallace). And it is indeed curious that but one out of 160-odd species of this diverse Australian family should have crossed the Wallace Line and established itself in Bali.

I am always astonished by these birds and by their abundance at higher elevations in Central Bali. I am also bewildered by their variable plumage which may be related as much to breeding condition as age and sex. We can be certain that those large-looking sunbirds, rifling nectar from the flowers with their long curved bills, are in fact Brown Honeyeaters. Some, however, are grey rather than brown, with varying amounts of olive-green wash on wings and tail. Some grey birds have black gapes (skin forming the corner of the mouth) and a pale patch behind the eye, and are presumed to be breeding males; browner birds with yellow gapes, yellow chins and throats, and no eye-patch are thought to be females or young birds. It is appropriate that the scientific name of this species is *indistincta*.

The voice is even more varied, a medley of sparrow chirps, bulbul chatter, shrike curses, and starling squeals. Also some quite pleasant whistled notes: *chew-whit*, a low urgent *shhh shhh*, and a hoarse double cough, as if the bird were clearing its throat. The species definitely calls for further study.

# Rainbow Lorikeet
*Trichoglossus haematodus*

Here we have another instance of an essentially Australian species crossing the Wallace Line to Bali, but failing to gain a foothold further west. It was formerly present in some numbers and, being mainly a blossom-feeder, may have invaded Bali on the introduction of coffee cultivation and the intensive planting of coral-bean (*Erythrina*) shade-trees which flower profusely. The unpalatable fact is, however, that there has been no official sighting of this bird since World War II.

As recently as three or four years ago, I used to see these Lorikeets, often in pairs, offered for sale in the Candikuning market-place at Bedugul. I was assured that they were of local provenance and were trapped high up in the mountains, which was doubt-

less true, for why would anyone bother to import birds to sell in Bedugul? But, try as I might, I could never find them in the wild. Now the cages are empty, which may be just as well.

These brilliant birds — never so brilliant in a cage —answered the description of the race *mitchelli* (found also on Lombok, but are they there yet?), with the blue head and belly plumes of Australian birds replaced by black. Possibly, like the Yellow-crested Cockatoos, formerly reported from Bali, periodic irruptions of eastern populations may see them reinstated here. But not for long I fear, if the trappers, supported by collectors, have their way.

# Yellow-throated Hanging-Parrot

*Loriculus pusillus*
CENTRE

Spending most of their lives upside-down, in which position they sleep, of exaggeratedly ungainly gait and exquisitely dainty feeding habits, these little comedians make marvellous pets.

In spite of the increased pressure of trapping, they somehow manage to persist in the wild, and I was recently elated to see a party of six or eight birds noisily alight high in the canopy of a giant fruiting *ficus* in forest by Lake Tamblingan (September 1988).

One of nine species of hanging-parrot inhabiting Indonesia, the Yellow-throated occurs only in Java and Bali. Note that the female lacks the yellow gorget. The Balinese name is *sririt*.

# Red-breasted Parakeet

*Psittacula alexandri*
TOP LEFT

In 1974 it was locally common on the northern forested slopes of the central mountain range. Trekking through isolated hamlets, the inhabitants would attempt to palm off a pair or more on you for a pittance. I have also seen odd birds in Sanur and adjacent mangrove, but these may have been escapes. Ranging from the Himalayas to Bali, this lovely parrot, violet-grey head separated from lavender-pink breast by a broad black moustache (whence its popular name, Moustached Parakeet), is not really a collector's item, so it should still be with us.

# Great-billed Parrot

*Tanygnathus megalorhynchos*
BOTTOM RIGHT

Confined to small islands, and the coasts of larger islands, in Indonesia, several authors have placed Bali within its area of distribution, but no authoritative source can be traced; the nearest published record of its occurrence is from Sumba. Frank Jarvis painted this magnificent parrot's portrait on the premises of a Singapore bird dealer.

# Chestnut-breasted Malkoha

*Phaenicophaeus curvirostris*

The Malkohas are large cuckoos, somewhat resembling the Coucals, that build their own nests and stay for the most part well-hidden in thick foliage, seldom taking flight or coming to the ground. But I have occasionally seen them in the air, gliding between trees, short wings and long tail spread in the shape of a Latin cross.

Several species inhabit Wallace's Oriental Region, being especially concentrated in Borneo and Sumatra, but the Chestnut-breasted is the only one that reaches Bali. And here it may be found fairly commonly in highland forest and low scrub

in the National Park. The challenge is to find it.

For all its size and arresting appearance, it is uncommonly difficult to observe. Usually only part of the bird is visible: the long tail, half glossy bottle-green, half burnished chestnut, suspended from a leafy bower; a glaring eye surrounded by a livid crimson patch and preceded by a broad pale green bill; a flash of chestnut breast. The effect is uncanny. Our eyes must be deceived.

Indeed the ruddy underparts, momentarily glimpsed, convince us that a squirrel, not a bird, is there. The vague brown form creeps through the canopy, reappearing now and then in gaps between the leaves, and still we are none the wiser. The only thing to do is clap or throw a stick, in the hope that whatever it is will erupt into view. But that is hardly fair; far better to wait quietly, or move on expectantly to some other spot.

Apart from the whoosh of its wings in flight, the Malkoha is a strangely silent bird. As an obvious target for the stealthy hunter equipped with lethal air-gun, it is well that it so remains. Birds in Borneo are said to emit a harsh chatter, like that of a magpie, or a low *kuk-kuk* whilst sitting in a tree. But I have never heard it make a sound.

I have read that Malkohas may also mewl and cluck and whine. And doubtless they have an infinitude of ways in which to express themselves to one another, more especially when assured that no human ear may hear them. Simply to behold them at liberty and at home in their green mansion is a thrill, and a stirring of the soul that cleaves to us forever.

It was not exactly the Chestnut-breasted Malkoha that Wordsworth had in view when he wrote:

> My dazzled sight he oft deceives
> A brother of the dancing leaves.

But it could well have been. It is a deception I would covet daily.

# CHECK-LIST OF THE BIRDS OF BALI

Every species known to have occurred in Bali is included under its respective family grouping. In each case the distinguishing scientific name is preceded by the established English one, though it will be noticed that certain of the latter have been changed to conform to recent proposals put forward by the Records Committee of the British Ornithologists' Union. Further, to enable bird-watchers to seek and observe these birds, a sequence of four letters accompanies each species according to the following key:

1. Most likely habitat in Bali
   O  Ocean, reef, lagoon, and rocky shore
   M  Mudflat, and adjacent mangrove swamp and fish-ponds
   C  Cultivation (including gardens), open and grassy tracts, with scattered trees, and intervening waterways
   F  Forest, forest edge and scrub (including overgrown ravines), and lakes with marshy fringes

2. Probable status
   R  Resident
   V  Visitor or passage migrant
   S  Straggler or irregular visitor

3. Incidence of occurrence
   A  Abundant
   C  Common (including locally common)
   U  Uncommon
   R  Rare

4. Best or most convenient place for viewing
   S  Sanur, through Suwung, to Kuta
   N  Nusa Dua, through Bukit, to Ulu Watu
   U  Ubud and environs
   B  Bedugul area
   W  West Bali and National Park
   P  Penida

An asterisk(*) following the code means the bird has not been seen in recent years and may possibly be extinct in Bali.

Family PROCELLARIIDAE: Petrels, Shearwaters
Bulwer's Petrel (*Bulweria bulwerii*)  O S U P

Family OCEANITIDAE: Storm-Petrels
Wilson's Storm-Petrel (*Oceanites oceanicus*)  O V C P
Swinhoe's Storm-Petrel (*Oceanodroma monorhis*)  O S R P

Family PODICIPEDIDAE: Grebes
Little Grebe (*Tachybaptus ruficollis*)  F R R B
Australian Little Grebe (*Tachybaptus novaehollandiae*)  M S R N

Family PHAETHONTIDAE: Tropicbirds
White-tailed Tropicbird (*Phaethon lepturus*)  O R U P

Family FREGATIDAE: Frigatebirds
Great Frigatebird (*Fregata minor*)  O V C S
Lesser Frigatebird (*Fregata ariel*)  O V C S
Christmas Frigatebird (*Fregata andrewsi*)  O S U S

Family PHALACROCORACIDAE: Cormorants
Little Pied Cormorant (*Phalacrocorax melanoleucos*)  M S R S
Little Black Cormorant (*Phalacrocorax sulcirostris*)  M S R S

Family ANHINGIDAE: Darters
Oriental Darter (*Anhinga melanogaster*)  F S R B*

Family SULIDAE: Boobies
Red-footed Booby (*Sula sula*)  O S R W
Brown Booby (*Sula leucogaster*)  O V C N

Family PELECANIDAE: Pelicans
Australian Pelican (*Pelecanus conspicillatus*)  F S U B
Great White Pelican (*Pelecanus onocrotalus*)  O S R W

Family ARDEIDAE: Herons, Egrets, Bitterns
Great-billed Heron (*Ardea sumatrana*)  O S R W
Grey Heron (*Ardea cinerea*)  M S U S
Purple Heron (*Ardea purpurea*)  M R C S
White-faced Heron (*Ardea novaehollandiae*)  O S R P
Great Egret(*Egretta alba*)  M S R S
Short-billed Egret (*Egretta intermedia*)  C R C U
Little Egret (*Egretta garzetta*)  C R C U
Pacific Reef-Egret (*Egretta sacra*)  O R C N
Cattle Egret (*Bubulcus ibis*)  C R C U
Javan Pond-Heron (*Ardeola speciosa*)  M R A S
Little Heron (*Butorides striatus*)  M R A S
Black-crowned Night-Heron (*Nycticorax nycticorax*)  M R U S
Yellow Bittern (*Ixobrychus sinensis*)  M V C S
Cinnamon Bittern (*Ixobrychus cinnamomeus*)  C R C U
Black Bittern (*Ixobrychus flavicollis*)  F S R B

Family CICONIIDAE: Storks
Milky Stork (*Mycteria cinerea*)  M S U S
Woolly-necked Stork (*Ciconia episcopus*)  M S U W
Lesser Adjutant (*Leptoptilus javanicus*)  M R C W

Family THRESKIORNITHIDAE: Ibises, Spoonbills
Glossy Ibis (*Plegadis falcinellus*)  M S R S
Royal Spoonbill (*Platalea regia*)  M S R S

Family ANATIDAE: Geese, Ducks
Diving Tree-Duck (*Dendrocygna arcuata*)  M S U S
Lesser Tree-Duck (*Dendrocygna javanica*)  M S U W
Grey Teal (*Anas gibberifrons*)  M R C S
Pacific Black Duck (*Anas superciliosa*)  M S R S*
Garganey (*Anas querquedula*)  O S U W

Family PANDIONIDAE: Osprey
Osprey (*Pandion haliaetus*)  O R U P

Family ACCIPITRIDAE: Hawks, Eagles
Crested Honey-Buzzard (*Pernis ptilorhynchus*)  F V C W
Black-shouldered Kite (*Elanus caeruleus*)  F S U W
Brahminy Kite (*Haliastur indus*)  M R C S
White-bellied Sea-Eagle (*Haliaeetus leucogaster*)  O R U N
Short-toed Eagle (*Circaetus gallicus*)  F S R W
Crested Serpent-Eagle (*Spilornis cheela*)  F R C B
Japanese Sparrowhawk (*Accipiter gularis*)  F V C W
Besra (*Accipiter virgatus*)  F S R B
Crested Goshawk (*Accipiter trivirgatus*)  F S R W
Chinese Goshawk (*Accipiter soloensis*)  F V C W

Gray-faced Buzzard (*Butastur indicus*) F V U W
Common Buzzard (*Buteo buteo*) F V R W
Black Eagle (*Ictinaetus malayensis*) F R U B
Booted Eagle (*Hieraaetus pennatus*) F V R W
Rufous-bellied Eagle (*Hieraaetus kienerii*) F V R W
Changeable Hawk-Eagle (*Spizaetus cirrhatus*) F S R W
Black-thighed Falconet (*Microhierax fringillarius*) F R U W
Spotted Kestrel (*Falco moluccensis*) C R C U
Oriental Hobby (*Falco severus*) F R R P
Peregrine Falcon (*Falco peregrinus*) O R R P

Family MEGAPODIIDAE: Megapodes
Orange-footed Scrubfowl (*Megapodius reinwardt*) F R R P*

Family PHASIANIDAE: Quail, Partridges, Pheasants
Blue-breasted Quail (*Coturnix chinensis*) C R R U
Red Junglefowl (*Gallus gallus*) F R U B
Green Junglefowl (*Gallus varius*) F R C W

Family TURNICIDAE: Button-Quail
Striped Button-Quail (*Turnix sylvatica*) C R R S
Barred Button-Quail (*Turnix suscitator*) C R C S

Family RALLIDAE: Rails, Crakes, Coots
Slaty-breasted Rail (*Gallirallus striatus*) M R R S
Red-legged Crake (*Rallina fasciata*) F S R W
Ruddy-breasted Crake (*Porzana fusca*) C R C U
White-browed Crake (*Porzana cinerea*) M R C S
Watercock (*Gallicrex cinerea*) C V C S
White-breasted Waterhen
(*Amaurornis phoenicurus*) C R C S
Common Moorhen (*Gallinula chloropus*) M R C S
Purple Swamphen (*Porphyrio porphyrio*) F S R B
Black Coot (*Fulica atra*) F S R B

Family JACANIDAE: Jacanas
Pheasant-tailed Jacana (*Hydrophasianus chirurgus*) F S R B

Family CHARADRIIDAE: Plovers
Grey Plover (*Pluvialis squatarola*) M V C S
Pacific Golden Plover (*Pluvialis fulva*) M V C S
Little Ringed Plover (*Charadrius dubius*) M V C S
Kentish Plover (*Charadrius alexandrinus*) M V C S
Malay Sand-Plover (*Charadrius peronii*) M V C S
Long-billed Plover (*Charadrius placidus*) M S R S
Mongolian Plover (*Charadrius mongolus*) M V A S
Greater Sand-Plover (*Charadrius leschenaultii*) M V C S
Oriental Plover (*Charadrius veredus*) C V U N

Family SCOLOPACIDAE: Curlews, Godwits, Sandpipers, Snipe
Western Curlew (*Numenius arquata*) M V U S
Whimbrel (*Numenius phaeopus*) M V C S
Eastern Curlew (*Numenius madagascariensis*) M V C S
Black-tailed Godwit (*Limosa limosa*) C V U S
Bar-tailed Godwit (*Limosa lapponica*) M V U S
Common Redshank (*Tringa totanus*) M V A S
Marsh Sandpiper (*Tringa stagnatilis*) M V U S
Common Greenshank (*Tringa nebularia*) M V C S
Green Sandpiper (*Tringa ochropus*) M S R S
Wood Sandpiper (*Tringa glareola*) C V A S
Terek Sandpiper (*Xenus cinereus*) M V R S
Common Sandpiper (*Actitis hypoleucos*) M V C S
Grey-tailed Tattler (*Heteroscelus brevipes*) M V C S
Ruddy Turnstone (*Arenaria interpres*) M V C S

Long-billed Dowitcher (*Limnodromus scolopaceus* M S R S
Pintail Snipe (*Gallinago stenura*) C V C S
Swinhoe's Snipe (*Gallinago megala*) C V U S
Red Knot (*Calidris canutus*) M V R S
Great Knot (*Calidris tenuirostris*) M V U S
Rufous-necked Stint (*Calidris ruficollis*) M V A S
Long-toed Stint (*Calidris subminuta*) C V A S
Sharp-tailed Sandpiper (*Calidris acuminata*) C V U S
Curlew Sandpiper (*Calidris ferruginea*) M V A S
Sanderling (*Crocethia alba*) M V C S
Ruff (*Philomachus pugnax*) C V R S

Family RECURVIROSTRIDAE: Stilts, Avocets
Black-winged Stilt (*Himantopus himantopus*) C S U S

Family PHALAROPODIDAE: Phalaropes
Red-necked Phalarope (*Phalaropus lobatus*) O V C P

Family BURHINIDAE: Thick-kness
Beach Thick-knee (*Esacus magnirostris*) O R R W

Family GLAREOLIDAE: Pratincoles
Oriental Pratincole (*Glareola maldiverum*) C V C S
Long-legged Pratincole (*Stiltia isabella*) C S R N

Family STERCORARIIDAE: Skuas
Pomarine Skua (*Stercorarius pomarinus*) O V U S
Arctic Skua (*Stercorarius parasiticus*) O V U S

Family LARIDAE: Gulls, Terns
Whiskered Tern (*Chlidonias hybridus*) C V C S
White-winged Black Tern (*Chlidonias leucopterus*) M V C S
Gull-billed Tern (*Gelochelidon nilotica*) M V C S
Common Tern (*Sterna hirundo*) M V C S
Arctic Tern (*Sterna paradisaea*) M V U S
Roseate Tern (*Sterna dougallii*) M V U S
Black-naped Tern (*Sterna sumatrana*) O V U S
Bridled Tern (*Sterna anaethetus*) O V U S
Little Tern (*Sterna albifrons*) M V C S
Greater Crested Tern (*Sterna bergii*) M V C S
Lesser Crested Tern (*Sterna bengalensis*) M V U S

Family COLUMBIDAE: Pigeons, Doves
Rock Pigeon (*Columba livia*) C R C P
Pink-necked Green-Pigeon (*Treron vernans*) C R C N
Orange-breasted Pigeon (*Treron bicincta*) F R C W
Grey-cheeked Green-Pigeon (*Treron griseicauda*) F R C B
Black-backed Fruit-Dove (*Ptilinopus cinctus*) F R R B
Pink-necked Fruit-Dove (*Ptilinopus porphyreus*) F R R B*
Black-naped Fruit-Dove (*Ptilinopus melanospila*) F R R W
Green Imperial Pigeon (*Ducula aenea*) F R U W
Pied Imperial Pigeon (*Ducula bicolor*) F R R W
Dark-backed Imperial Pigeon (*Ducula lacernulata*) F R U B
Little Cuckoo-Dove (*Macropygia ruficeps*) F R U B
Indonesian Cuckoo-Dove (*Macropygia emiliana*) F R U B
Spotted Dove (*Streptopelia chinensis*) C R A S
Island Turtle-Dove (*Streptopelia bitorquata*) M R C S
Zebra Dove (*Geopelia striata*) C R U S
Green-winged Pigeon (*Chalcophaps indica*) F R C W

Family PSITTACIDAE: Parrots
Yellow-crested Cockatoo (*Cacatua sulphurea*) C S R P
Rainbow Lorikeet (*Trichoglossus haematodus*) F R R B
Great-billed Parrot (*Tanygnathus megalorynchos*) F S R P*
Red-breasted Parakeet (*Psittacula alexandri*) F R U B

Yellow-throated Hanging-Parrot
(*Loriculus pusillus*)                                         F R U B

Family CUCULIDAE: Cuckoos
Large Hawk-Cuckoo (*Cuculus sparverioides*)     F S R U
Hodgson's Hawk-Cuckoo (*Cuculus fugax*)         F S R W
Oriental Cuckoo (*Cuculus saturatus*)           F V C B
Banded Bay Cuckoo (*Cacomantis sonneratii*)     F S R W
Plaintive Cuckoo (*Cuculus merulinus*)          C R C S
Indonesian Brush Cuckoo (*Cuculus sepulcralis*) F R C B
Horsfield's Bronze-Cuckoo (*Chrysococcyx basalis*) C V C S
Drongo Cuckoo (*Surniculus lugubris*)           F R U W
Common Koel (*Eudynamys scolopacea*)            C R U U
Chestnut-breasted Malkoha
(*Phaenicophaeus curvirostris*)                 F R U B
Greater Coucal (*Centropus sinensis*)           C R C U
Lesser Coucal (*Centropus bengalensis*)         C R C U

Families TYTONIDAE and STRIGIDAE: Owls
Barn Owl (*Tyto alba*)                          C R R S
Bay Owl (*Phodilus badius*)                     F R R W
Collared Scops-Owl (*Otus bakkamoena*)          C R C U
Barred Eagle-Owl (*Bubo sumatranus*)            F R R W
Buffy Fish-Owl (*Ketupa ketupu*)                F R R W
Javan Owlet (*Glaucidium castanopterum*)        F R R W
Brown Hawk-Owl (*Ninox scutulata*)              F R R W

Family CAPRIMULGIDAE: Nightjars
Large-tailed Lightjar (*Caprimulgus macrurus*)  F R U W
Savannah Nightjar (*Caprimulgus affinis*)       C R C S

Family HEMIPROCNIDAE: Tree-Swifts
Grey-rumped Tree-Swift (*Hemiprocne longipennis*) F R C W

Family APODIDAE: Swifts
White-bellied Swiftlet (*Collocalia esculnta*)  C R A S
Edible-nest Swiftlet (*Aerodramus fuciphagus*)  C R C U
Mossy-nest Swiftlet (*Aerodramus salangana*)    F R C B
White-throated Needletail
(*Hirundapus caudacutus*)                       F V U W
White-vented Needletail
(*Hirundapus cochinchinensis*)                  F S U W
Brown Needletail (*Hirundapus giganteus*)       C R U U
Fork-tailed Swift (*Apus pacificus*)            C V C U
House Swift (*Apus affinis*)                     C R C N
Asian Palm Swift (*Cypsiurus balasiensis*)      C R C S

Family ALCEDINIDAE: Kingfishers
River Kingfisher (*Alcedo atthis*)              C S R W*
Blue-eared Kingfisher (*Alcedo meninting*)      F R R W*
Small Blue Kingfisher (*Alcedo coerulescens*)   M R C S
Rufous-backed Kingfisher (*Ceyx rufidorsus*)    F R U W
Stork-billed Kingfisher (*Halcyon capensis*)    M R R W
Collared Kingfisher (*Halcyon chloris*)         C R C S
Java Kingfisher (*Halcyon cyanoventris*)        C R C U
Scared Kingfisher (*Halcyon sancta*)            M V C S

Family MEROPIDAE: Bee-eaters
Bay-headed Bee-eater (*Merops leschenaulti*)    F R C W
Blue-tailed Bee-eater (*Merops philippinus*)    C V C S
Rainbow Bee-eater (*Merops ornatus*)            C S R S

Family CORACIIDAE: Rollers
Dollarbird (*Eurystomus orientalis*)            F R C W
Family BUCEROTIDAE: Hornbills

Wreathed Hornbill (*Rhyticeros undulatus*)      F R C W
Southern Pied Hornbill (*Anthracoceros convexus*) F R U W

Family CAPITONIDAE: Barbets
Lineated Barbet (*Megalaima lineata*)           F R C W
Blue-crowned Barbet (*Megalaima armillaris*)    F R C B
Blue-eared Barbet (*Megalaima australis*)       F R C B
Coppersmith Barbet (*Megalaima haemacephala*)   C R C U

Familly PICIDAE: Woodpeckers
Laced Woodpecker (*Picus vittatus*)             F R U W
Common Golden-backed Woodpecker
(*Dinopium javanense*)                          C R U U
Greater Golden-backed Woodpecker
(*Chrysocolaptes lucidus*)                      F R U W
White-bellied Black Woodpecker
(*Dryocopus javensis*)                          F R R W*
Fulvous-breasted Woodpecker (*Picoides macei*)  C R C U
Brown-capped Woodpecker (*Picoides moluccensis*) M R R S

Family PITTIDAE: Pittas
Banded Pitta (*Pitta guajana*)                  F R U W
Elegant Pitta (*Pitta elegans*)                 F R R P*

Family ALAUDIDAE: Larks
Singing Bush Lark(*Mirafra javanica*)           C R C S

Family HIRUNDINIDAE: Swallows
Barn Swallow (*Hirundo rustica*)                C V A S
Pacific Swallow (*Hirundo tahitica*)            C R A S
Red-rumped Swallow (*Hirundo striolata*)        C V C U
Asian House Martin (*Delichon dasypus*)         F V R W

Family CAMPEPHAGIDAE: Cuckoo-Shrikes, Minivets
Black-winged Flycatcher-Shrike
(*Hemipus hirundinaceus*)                       F R C B
Large Cuckoo-Shrike (*Coracina novaehollandiae*) F R U B
Lesser Cukoo-Shrike (*Coracina fimbriata*)      F R C B
White-winged Triller (*Lalage sueurii*)         M R C S
Small Minivet (*Pericrocotus cinnamomeus*)      F R C W
Scarlet Minvert (*Pericrocotus flammeus*)       F R C B

Family CHILOROPSEIDAE: Ioras, Leafbirds
Common Iora (*Aegithina tiphia*)                C R C S

Family PYCNONOTIDAE: Bulbuls
Black-headed Bulbul (*Pycnonotus atriceps*)     F R U W
Black-crested Bulbul (*Pycnonotus melanicterus*) F R U W
Sooty-headed Bulbul (*Pycnonotus aurigaster*)   C R C S
Yellow-vented Bulbul (*Pycnonotus goiavier*)    C R A S
Orange-spotted Bulbul (*Pycnonotus bimaculatus*) F R C B
Grey-cheeked Bulbul (*Criniger bres*)           F R C B

Family DICRURIDAE: Drongos
Black Drongo (*Dicrurus macrocercus*)           C R C N
Ashy Drongo (*Dicrurus leucophaeus*)            F R C B
Hair-crested Drongo (*Dicrurus hottentottus*)   F R C W
Greater Racket-tailed Drongo
(*Dicrurus paradiseus*)                         F R U B

Family ORIOLIDAE: Orioles
Black-naped Oriole (*Oriolus chinensis*)        C R C S

Family CORVIDAE: Jays, Magpies, Crows
Racket-tailed Treepie (*Crypsirina temia*)      F R U W

Slender-billed Crow (*Corvus enca*)    F R U W
Large-billed Crow (*Corvus macrorhynchos*)    C R C S

Family PARIDAE: Tits
Great Tit (*Parus major*)    F R C B

Family TIMALIIDAE: Babblers
Horsfield's Babbler (*Trichastoma sepiarium*)    F R C U
Chestnut-backed Scimitar-Babbler
  (*Pomatorhinus montanus*)    F R C B
Pearl-cheeked Babbler (*Stachyris melanothorax*)    F R C B

Family TURDIDAE: Thrushes
Lesser Shortwing (*Brachypteryx leucophrys*)    F R C B
Magpie Robin (*Copsychus saularis*)    C R C S
White-crowned Forktail (*Enicurus leschenaulti*)    F R C U
Pied Bushchat (*Saxicola caprata*)    C R C U
Sunda Whistling Thrush (*Myophonus glaucinus*)    F R R B
Orange-headed Thrush (*Zoothera citrina*)    F R U B
Siberian Thrush (*Zoothera sibirica*)    F S U B
Scaly Thrush (*Zoothera dauma*)    F R R B
Sunda Ground Thrush (*Zoothera andromedae*)    F R R B
Eye-browed Thrush (*Turdus obscurus*)    F S U B

Family ACANTHIZIDAE: Australian Warblers
Golden-bellied Flyeater (*Gerygone sulphurea*)    M R C S

Family SYLVIIDAE : Old World Warblers
Sunda Flycatcher-Warbler (*Seicercus grammiceps*)    F R C B
Yellow-bellied Warbler (*Abroscopus superciliaris*)    F R U W
Arctic Warbler (*Phylloscopus borealis*)    F V C W
Mountain Leaf-Warbler (*Phylloscopus trivirgatus*)    F R C B
Eastern Great Reed-Warbler
  (*Acrocephalus orientalis*)    C V U S
Pallas' Grasshopper-Warbler (*Locustella certhiola*)    F V U B
Striated Warbler (*Megalurus palustris*)    C R C N
Ashy Tailorbird (*Orthotomus sepium*)    C R C S
Mountain Tailorbird (*Orthotomus cuculatus*)    F R C B
Bar-winged Prinia (*Prinia familiaris*)    C R A S
Zitting Cisticola (*Cisticola juncidis*)    C R A U
Golden-headed Cisticola (*Cisticola exilis*)    C R A U
Indonesian Bush-Warbler (*Cettia vulcania*)    F R U B
Russet Bush-Warbler (*Bradypterus seebohmi*)    F R R B

Family MUSCICAPIDAE: Flycatchers
Fulvous-chested Flycatcher (*Rhinomyias olivacea*)    F R U B
Asian Brown Flycatcher (*Muscicapa latirostris*)    F V U W
Yellow-rumped Flycatcher (*Ficedula zanthopygia*)    F V U W
Mugimaki Flycatcher (*Ficedula mugimaki*)    F V R B
Snowy-browed Flycatcher (*Ficedula hyperythra*)    F R U B
Little Pied Flycatcher (*Ficedula westermanni*)    F R C B
Mangrove Blue Flycatcher (*Cyornis rufigastra*)    M S R W
Grey-headed Flycatcher (*Culicicapa ceylonensis*)    F R C B
Pied Fantail (*Rhipidura javanica*)    C R C S
Black-naped Monarch (*Hypothymis azurea*)    F R C B

Family PACHYCEPHALIDAE: Whistlers
Mangrove Whistler (*Pachycephala grisola*)    M R U W
Golden Whistler (*Pachycephala pectoralis*)    F R C B

Family MOTACILLIDAE: Wagtails, Pipits
Grey Wagtail (*Motacilla cinerea*)    C V U U
Yellow Wagtail (*Motacilla flava*)    C V C S
Common Pipit (*Anthus novaeseelandiae*)    C R C N

Family ARTAMIDAE: Woodswallows
White-breasted Woodswallow
  (*Artamus leucorynchus*)    C R C S
Family LANIIDAE: Shrikes
Brown Shrike (*Lanius cristatus*)    C V U S
Tiger Shrike (*Lanius tigrinus*)    F S R N
Long-tailed Shrike (*Lanius schach*)    C R C S

Family STURNIDAE: Starlings, Mynas
Philippine Glossy Starling (*Aplonis panayensis*)    C R C U
Short-tailed Glossy Starling (*Aplonis minor*)    F S U B
Asian Pied Starling (*Sturnus contra*)    C R C N
Black-winged Starling (*Sturnus melanopterus*)    C R C N
Bali Starling (*Leucopsar rothschildi*)    F R R W
White-vented Myna (*Acridotheres javanicus*)    C R C S
Hill Myna (*Gracula religiosa*)    F R U W

Family NECTARINIIDAE: Sunbirds, Spiderhunters
Brown-throated Sunbird (*Anthreptes malacensis*)    C R C S
Olive-backed Sunbird (*Nectarinia jugularis*)    C R A S
Little Spiderhunter (*Arachnothera longirostra*)    C R C U
Grey-breasted Spiderhunter (*Arachnothera affinis*)    F R R B

Family MELIPHAGIDAE: Honeyeaters
Brown Honeyeater (*Lichmera indistincta*)    C R C B

Family DICAEIDAE: Flowerpeckers
Yellow-vented Flowerpecker
  (*Dicaeum chrysorrheum*)    F R U W
Orange-bellied Flowerpecker
  (*Dicaeum trigonostigma*)    F R U B
Plain Flowerpecker (*Dicaeum concolor*)    F R U W
Scarlet-headed Flowerpecker (*Dicaeum trochileum*)    C R C S
Blood-breasted Flowerpecker
  (*Dicaeum sanguinolentum*)    F R C B
Red-chested Flowerpecker (*Dicaeum maugei*)    C R C P

Family ZOSTEROPIDAE: White-eyes
Oriental White-eye (*Zosterops palpebrosus*)    C R C U
Mountain White-eye (*Zosterops montanus*)    F R C B
Mangrove White-eye (*Zosterops chloris*)    M R C W
Javan Grey-throated White-eye
  (*Lophozosterops javanicus*)    F R A B

Family PLOCEIDAE: Sparrows, Weavers
Eurasian Tree Sparrow (*Passer montanus*)    C R A S
Baya Weaver (*Ploceus philippinus*)    C R R S
Streaked Weaver (*Ploceus manyar*)    C R A S

Family ESTRILDIDAE: Waxbills, Munias
Strawberry Waxbill (*Amandava amandava*)    C R R U
Java Sparrow (*Padda oryzivora*)    C R C S
Javan Munia (*Lonchura leucogastroides*)    C R A S
Scaly-breasted Munia (*Lonchura punctulata*)    C R A S
Chestnut Munia (*Lonchura malacca*)    C R C U
White-headed Munia (*Lonchura maja*)    C R C S
Black-faced Munia (*Lonchura molucca*)    C R C P

# LIST OF PROBABLES

The following birds may be expected to turn up in Bali , probable observations having been made of many as noted, and it would be good to have confirmation of their existence here for eventual publication:

Wedge-tailed Shearwater, *Puffinus pacificus* —— Bali Strait (Sep)
Red-billed Tropicbird, *Phaethon aethereus* —— Buleleng (Sep)
Little Cormorant, *Phalacrocorax niger* —— Pesanggaran (Sep)
Chinese Egret, *Egretta eulophetes* —— Pesanggaran (May)
Nankeen Night-Heron, *Nycticorax caledonicus*
Black Kite, *Milvus migrans* —— Gitgit (Mar)
Shikra Goshawk, *Accipiter badius* —— Petitenget (Feb)
Nankeen Kestrel, *Falco cenchroides*
Northern Hobby, *Falco subbuteo*
Horsfield's Hill Partridge, *Arborophila orientalis* —— Tamblingan (Aug)
Dusky Moorhen, *Gallinula tenebrosa* —— Lake Buyan (Aug)
Greater Painted-Snipe, *Rostratulata benghalensis*
Little Curlew, *Numenius minutus* —— Suwung (Jan)
Snipe-billed Dowitcher, *Limnodromus semipalmatus* —— Mertha Sari (Dec)
Broad-billed Sandpiper, *Limicola falcinellus*
Long-tailed Skua, *Stercorarius longicaudus* —— Legian (Oct)
Black-headed Gull, *Larus ridibundus*
Chinese Crested Tern, *Sterna bernsteini* —— Sanur (Mar)
Brown Noddy, *Anous stolidus*
White-capped Noddy, *Anous minutus* —— Lembongan and north coast (Sep)
Nicobar Pigeon, *Caloenas nicobarica* —— (extinct?)
Greater Green Leafbird, *Chloropsis sonnerati* —— Wangaya Gedé (Aug)
Chestnut-capped Babbler, *Timalia pileata* —— Mt. Catur Bedugul (Dec)
Temminck's Jungle Babbler, *Trichastoma pyrrhogenys*
Javan Fulvetta, *Alcippe pyrrhoptera* —— Bali Barat (Nov)
Chestnut-capped Thrush, *zoothera interpres*
Eastern Crowned Warbler, *Phylloscopus coronatus* —— Bedugul (Dec)
Clamorous Reed-Warbler, *Acrocephalus stentoreus* —— Lake Buyan (Dec)
Rufous-chested Flycatcher, *Ficedula dumetoria*
Tawny-breasted Parrot-Finch, *Erythrura hyperythra*

I shall be gald to receive and acknowledge all reports of sightings of the above birds in Bali— as well as any observations on the Bali birds in general — either c/o the Publishers, Periplus Editions (HK) Pte Ltd, or at P.O. Box 400, Denpasar Bali 80001, Indonesia.

—*Victor Mason*